UTAH
WILDLIFE VIEWING GUIDE

Jim Cole

FALCON PRESS

D1279968

ACKNOWLEDGMENTS

Dale Bosworth, Wasatch-Cache Forest Supervisor, in his strong commitment to the Watchable Wildlife Program, provided funding for salary, travel, and incidental costs for the project manager position. Tim Provan, Director of the Utah Division of Wildlife Resources, and Reed Stalder, through BLM State Director James Parker, have been early and strong supporters of the project.

The Core Group, which directed the development of the Guide, included Jim Cole, Project Manager, Wasatch-Cache National Forest; Randy Radant, Utah Division of Wildlife Resources; Terry Cole and Jerry Farringer, Bureau of Land Management; Bob Benton, U.S. Fish and Wildlife Service; Dennis Burns, Utah Division of Parks and Recreation; Jamie Gentry, Zion Natural History Association; Stacey Clark, Utah Travel Council; and Jim Naegle, Utah Department of Transportation.

Sara Vickerman, who consulted in behalf of the Defenders of Wildlife Viewing Guide series, provided critical guidance throughout the development of this guide. Rene´ Eisenbart designed and developed the ecosystem artwork.

Don Paul, Utah Division of Wildlife Resources, provided ecological information for the Great Salt Lake ecosystem artwork. Andy White, Past-President of the Utah Audubon Society, provided the unique information on scientific name translations.

Governor Norman Bangerter, through his administrative assistant Curt Garner, also provided encouragement and support.

Others who contributed to the project included Susan Giannettino, Jim Mower, Garth Heaton, Bill Herr, Catherine Quinn, John Knudson, Gene Findlay, Bob Ruesink, Penny Thomas, Steve Howe, Doug Thompson, Allen Stokes, Wes Shields, Heather Musclow, Geoff Walsh, Bob Hurley, Amos Wright, Jim Burruss, and the many dedicated individuals who nominated viewing sites and reviewed the manuscript.

Author
Jim Cole
Wildlife Biologist,
Wasatch-Cache National Forest

Project Consultant
Sara Vickerman
Regional Program Director,
Defenders of Wildlife

Design, typesetting, and other prepress work by Falcon Press, Helena, Montana
Printed in Malaysia

Library of Congress Number 90-080041
ISBN 1-56044-023-6

Front cover photo: Long-tailed weasel by Jeffrey T. Hogan
Artwork on pages 10-15 by René Eisenbart

CONTENTS

The stealthy cougar is common to almost all habitats in Utah, but it favors rugged terrain where prey species are abundant. It is rarely seen because it is mainly active at night. JAN L. WASSINK

UTAH WILDLIFE VIEWING GUIDE
PROJECT SPONSORS

The FOREST SERVICE, U.S. DEPARTMENT OF AGRICULTURE, has a mandate to protect, improve, and wisely use the nation's forest and range resources for multiple purposes to benefit all Americans. The six national forests of Utah are sponsors of this program to promote awareness and enjoyment of fish and wildlife on our national forest system lands. USDA Forest Service, 324 25th Street, Ogden, UT 84401. (801) 625-5347.

The BUREAU OF LAND MANAGEMENT is responsible for the balanced management of the public lands and resources and their various values so that they are considered in a combination that will best serve the needs of the American people. Management is based upon the principles of multiple use and sustained yield, in a combination of uses that takes into account the long-term needs of future generations for renewable and nonrenewable resources. Utah State Office, Bureau of Land Management, 324 South State Street, Salt Lake City, UT 84111. (801) 539-4001.

The UTAH DIVISION OF WILDLIFE RESOURCES, as the wildlife authority of the state, works to assure the future of protected wildlife for its intrinsic, scientific, educational, and recreational values through protection, propagation, management, conservation, and distribution of protected wildlife throughout the state. In recognition of the diverse public interests in Utah's wildlife resources, the Division is sponsoring this program to increase awareness and support for wildlife values and conservation programs. Utah Division of Wildlife Resources, 1596 West North Temple, Salt Lake City, UT 84116. (801) 596-8660.

The UTAH DEPARTMENT OF TRANSPORTATION is responsible for providing safe, reliable, and well maintained highways throughout Utah. Providing access to the State's recreational and scenic areas, and preserving the environment are among the Department's highest priorities. The Department also publishes the official highway map, which is available free-of-charge at welcome centers and visitor information offices across the State. Utah Department of Transportation, 4501 South 2700 West, Salt Lake City, UT 84119. (801) 965-4104.

DEFENDERS OF WILDLIFE is a national, nonprofit organization of more than 80,000 members and supporters dedicated to preserving the natural abundance and diversity of wildlife and its habitat. A one-year membership is $20 and includes six issues of the bimonthly magazine, *Defenders*. To join or for further information, write or call Defenders of Wildlife, 1244 Nineteenth Street, N.W., Washington, DC 20036. (202) 659-9510.

The U.S. FISH AND WILDLIFE SERVICE is pleased to support the Watchable Wildlife effort in furtherance of its mission to preserve, protect, and enhance fish and wildlife resources and their habitats for the use and enjoyment by the American public. For more information contact U.S. Fish and Wildlife Service, 1745 West 1700 South, Salt Lake City, UT 84104. (801) 524-5630.

The UTAH DIVISION OF PARKS AND RECREATION is the recreation authority for the State of Utah. This includes the management, protection, and conservation of a number of State Park areas. Historical parks, natural areas, and recreation parks are part of the system. The Division also administers other recreational programs including the Comprehensive Outdoor Recreation Plan, boating, and off-highway vehicle programs. Utah Division of Parks and Recreation, 1636 West North Temple, Salt Lake City, UT 84116. (801) 538-7220.

STATE OF UTAH
OFFICE OF THE GOVERNOR
SALT LAKE CITY
84114

NORMAN H. BANGERTER
GOVERNOR

Wildlife is a significant part of the Utah experience! Whether we visit Canyonlands with its desert bighorn sheep, or the High Uintas with its pine marten, ptarmigan, or mountain goat, or the Henry Mountains and its bison herd, or the marshes of the Great Salt and its abundance of waterfowl and shorebirds, or....well soon you get the picture. Utah's wildlife resources are as diverse as its seemingly endless habitat.

Utah is famous for its hunting recreation, particularly mule deer hunting. But almost undiscovered is perhaps an even greater opportunity to watch or photograph wildlife, to simply enjoy and experience wildlife in Utah's abundant wildlands.

This is my invitation to Utah's citizens and visitors alike to use and enjoy this book as you explore the mountains and deserts of Utah to experience its wildlife. A partnership of federal and state wildlife and land management agencies, coupled with conservation organizations, has developed this viewing guide to help you enjoy and learn more about our wildlife. About ninety sites throughout the State are identified in the Guide and marked with highway signs at each location.

Enjoy!

Sincerely,

Norman H. Bangerter
Governor

THE WATCHABLE WILDLIFE PROGRAM

Historically, wildlife management in Utah has largely focused on hunted species. Sportsmen, through license fees and federal taxation on firearms and fishing tackle, have carried the burden of funding state wildlife programs in the past and this remains largely the case today. Often, these programs have benefited non-hunted species of wildlife through habitat enhancement and preservation. Direct management for appreciative uses has been a low-key effort by comparison.

Recently, public interest in wildlife-oriented activities such as photography and viewing have taken a dramatic upturn. In response to this situation, federal and state land and wildlife management agencies in Utah have formed a partnership with conservation organizations. This effort, known as the Watchable Wildlife Program, provides Utah's citizens and visitors alike a new opportunity to enjoy and better appreciate one of its greatest heritages—its wildlife. The Utah Wildlife Viewing Guide is the first step in this effort and will serve as the focal point for a complete Watchable Wildlife Program in the years to come. The next step, enhancement of individual viewing sites, is just getting under way in the state. Site enhancement involves such things as interpretive signing, trail development, construction of viewing blinds or platforms, and provision of parking and restroom facilities.

An additional benefit of the Watchable Wildlife Program is to expand our understanding of the world around us. It can help us to better value diversity or to see the value of all components of our natural environment. Species that are neither hunted, viewed, nor photographed may play a critical role in the health of another species, or indeed, an entire ecosystem. By viewing wildlife in its natural habitat, perhaps we may better appreciate the importance of all elements of the biological and physical world around us.

The largest common owl in Utah, these great horned owl nestlings will grow to have a wingspan of about four feet. When grown, they will use their superior eyesight to hunt for a variety of rabbits, rodents, and birds. J. KIRK GARDNER

VIEWING HINTS

• The first and last hours of daylight are generally the best times to view or photograph most species. Seasonally, spring and early summer are the best times to view many species such as songbirds, small mammals, and hoofed mammals since they are most active throughout the day during this period.

• Be quiet. Quick movements and loud noises will normally scare wildlife. Since your car or boat is a good viewing blind, you may actually see more by remaining in the vehicle. Streams should be approached slowly and vegetation used as a screen to avoid scaring fish in shallow water. Notice how much more often you see animals when you are still than when you are moving. Whisper when you speak.

• Use as many viewing aids as possible. Binoculars or spotting scopes are always desirable to enhance your observations. Field guides are helpful with identification and other pertinent facts. Polarized glasses help reduce glare and make fish viewing easier.

• Be patient. Wait quietly for animals to enter or return to an area. Give yourself enough time to allow animals to move within your view. Patience is usually rewarded with a more complete wildlife experience.

OUTDOOR ETHICS

• Honor the rights of private landowners. Gain permission of private landowners before entering their property.

• Honor wildlife's requirement of free movement. Feeding, touching, or otherwise harassing wildlife is inappropriate. Young wild animals that appear to be alone have not been abandoned; allow them to find their own way.

• Honor the rights of others to enjoy their viewing experience. Loud noises, quick movements, or extraordinary behavior that might scare wildlife is inappropriate. Wait your turn or seek another viewing opportunity.

• Honor your own right to enjoy the outdoors in the future. Leave wildlife habitat in better condition than you found it. Pick up litter that you might encounter at a viewing site and dispose of it properly.

MAP INFORMATION

Utah is divided into the nine travel
regions shown on this map. Wildlife
viewing sites are numbered consecutively
in a general pattern from north to south.
Each region forms a separate section in
this book, and each section begins with
a detailed map of that region.

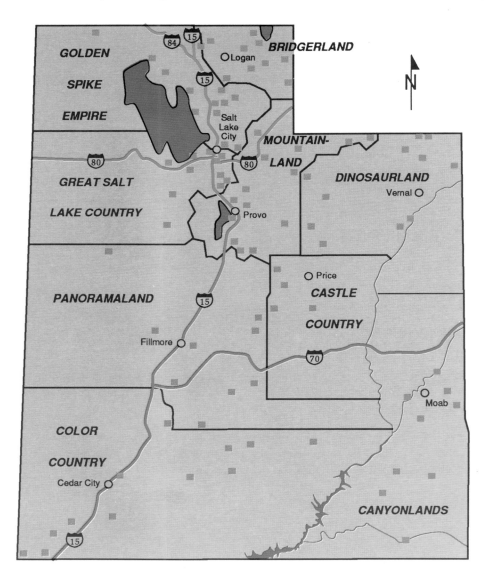

HOW TO USE THIS GUIDE

This Guide is arranged to coincide with the nine travel regions of the state, with each viewing site described in the appropriate travel region.

Each site description includes the featured wildlife and the habitat in which each species may be seen. Additional information relating to viewing probability (high, moderate, limited, rare), viewing season, and specific directions to the site is also provided. NOTES OF PRECAUTION RELATING TO ROAD CONDITIONS, SAFETY, VIEWING LIMITATIONS, AND LAND OWNERSHIP RESTRICTIONS ARE NOTED IN CAPITAL LETTERS.

Each site description also displays symbols for featured wildlife, available facilities, site owner/manager initials, plus a telephone number where additional site information may be obtained. Each viewing site will be marked by the binoculars logo, symbolic of the wildlife viewing sites across the nation.

FEATURED WILDLIFE

Carnivores   Hoofed mammals   Small mammals   Freshwater mammals   Waterfowl   Upland birds   Songbirds

Birds of prey   Fish   Wildflowers   Reptiles, amphibians   Shorebirds   Wading birds

HIGHWAY SIGNS

As you travel across Utah, look for these special highway signs that identify wildlife viewing sites. Most signs show the binoculars logo or the words "Wildlife Viewing Area," with an arrow pointing toward the site.

SITE OWNER/MANAGER ABBREVIATIONS

USFS U.S. Forest Service
BLM Bureau of Land Management
UDWR Utah Division of Wildlife Resources
UDPR Utah Division of Parks and Recreation

NPS National Park Service
USFWS U.S. Fish and Wildlife Service
PVT Private ownership
NP Nonprofit organizations

FACILITIES AND RECREATION

P Parking   🚶 Hiking   ▲ Campground   ♿ Handicap accessible

$ Entry fee   Small boats   🪑 Picnic

 Boat ramp   Restrooms   Restaurant   Lodging

9

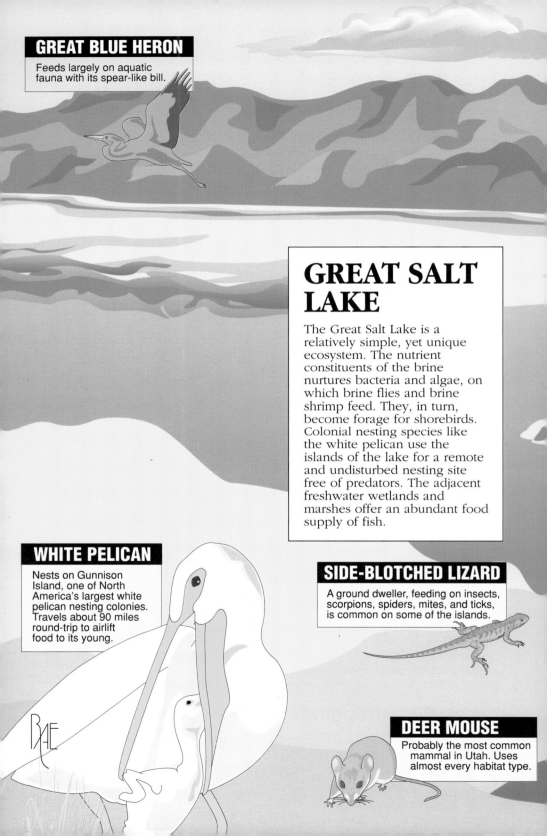

GREAT BLUE HERON
Feeds largely on aquatic fauna with its spear-like bill.

GREAT SALT LAKE

The Great Salt Lake is a relatively simple, yet unique ecosystem. The nutrient constituents of the brine nurtures bacteria and algae, on which brine flies and brine shrimp feed. They, in turn, become forage for shorebirds. Colonial nesting species like the white pelican use the islands of the lake for a remote and undisturbed nesting site free of predators. The adjacent freshwater wetlands and marshes offer an abundant food supply of fish.

WHITE PELICAN
Nests on Gunnison Island, one of North America's largest white pelican nesting colonies. Travels about 90 miles round-trip to airlift food to its young.

SIDE-BLOTCHED LIZARD
A ground dweller, feeding on insects, scorpions, spiders, mites, and ticks, is common on some of the islands.

DEER MOUSE
Probably the most common mammal in Utah. Uses almost every habitat type.

CALIFORNIA GULL

Some 80,000 nest and feed on a variety of food sources, including brine shrimp, reptiles, and small mammals.

BRINE FLY

A huge biomass of this swarming, non-biting insect provides food for migratory shorebirds.

GREEN ALGAE

Multiplies rapidly by feeding on brine nutrients. In shallow areas the lake takes on the algae's color.

PINTAIL

Nests in the east-side marshes of the lake. Feeds on aquatic plants, seeds, and small animals.

BRINE SHRIMP

Feed on green algae, often consuming so much algae in late summer that algae-tinted water begins to clear.

WILSON'S PHALAROPE

Arrives from breeding grounds in mid-June and leaves for wintering habitat about two months later, often weighing up to 75 percent more.

EARED GREBE

Nests on mats of floating vegetation in marshes.

BEAVER

Dam building maintains a high water table in meadows, creating habitat for vegetation such as willows and sedges.

NORTHERN HARRIER

Hunts prey by gliding slowly at low elevations over meadows and grasslands. Nests on ground in shrubby areas near marshes.

WHITE-CROWNED SPARROW

Lives in the low-shrub thickets associated with riparian zones. Feeds largely on seeds and insects.

WILLOW SHRUB

Occupies areas with high water tables. Root structure stabilizes the stream bank, while the crown shades the stream, cooling the water to benefit fish.

CUTTHROAT TROUT

Native species feeds on aquatic insects in the stream, and terrestial insects that fall into the water from streamside shrubs.

TIGER SALAMANDE

Dwells in rotten logs, animal burrows, or other underground moist places. Serves as prey for various birds and mammals.

ASPEN

Sometimes associated with riparian zones, aspen provides food (bark) for beaver as well as structure (limbs) for beaver dams.

RIPARIAN ZONE

Riparian areas are the lush, green vegetation zones along creeks and rivers, around seeps, springs, lakes, and reservoirs, and in bogs and wet meadows. They are among the most productive habitats despite their relatively minor proportional occurrence. Some 75 percent of all vertebrates are dependent on the unique and diverse habitat niches found in riparian areas for at least a portion of their life requirements.

MOOSE

Forages on willows, other browse species, and aquatic vegetation in beaver ponds or lakes.

LONG-TAIL VOLE

Feeds on meadow grasses and serves as prey for predatory mammals and birds.

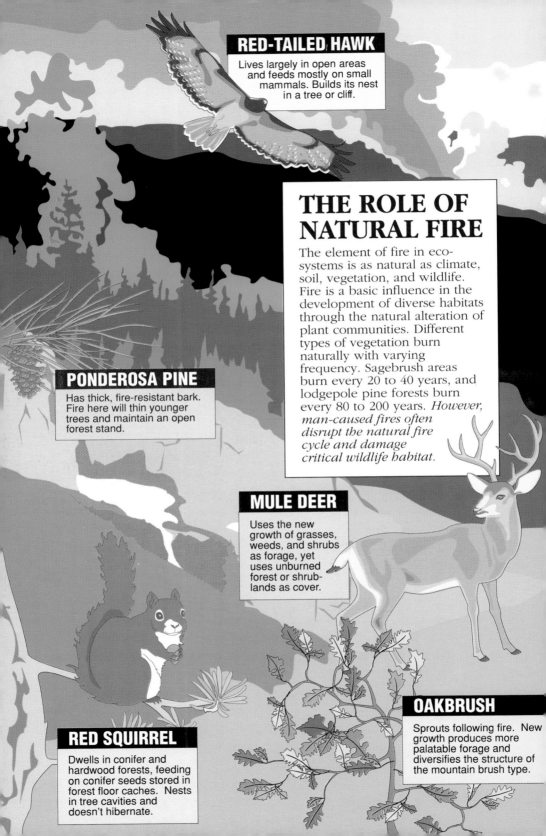

RED-TAILED HAWK
Lives largely in open areas and feeds mostly on small mammals. Builds its nest in a tree or cliff.

THE ROLE OF NATURAL FIRE
The element of fire in eco-systems is as natural as climate, soil, vegetation, and wildlife. Fire is a basic influence in the development of diverse habitats through the natural alteration of plant communities. Different types of vegetation burn naturally with varying frequency. Sagebrush areas burn every 20 to 40 years, and lodgepole pine forests burn every 80 to 200 years. *However, man-caused fires often disrupt the natural fire cycle and damage critical wildlife habitat.*

PONDEROSA PINE
Has thick, fire-resistant bark. Fire here will thin younger trees and maintain an open forest stand.

MULE DEER
Uses the new growth of grasses, weeds, and shrubs as forage, yet uses unburned forest or shrub-lands as cover.

OAKBRUSH
Sprouts following fire. New growth produces more palatable forage and diversifies the structure of the mountain brush type.

RED SQUIRREL
Dwells in conifer and hardwood forests, feeding on conifer seeds stored in forest floor caches. Nests in tree cavities and doesn't hibernate.

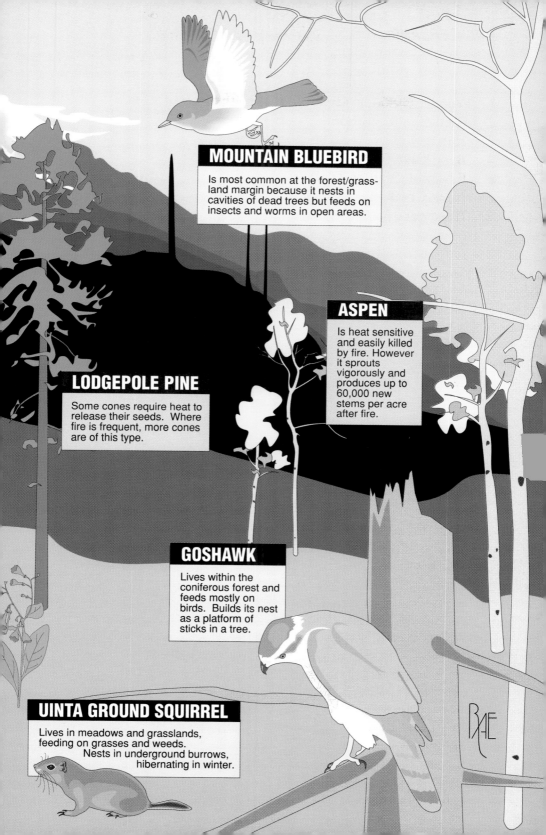

MOUNTAIN BLUEBIRD

Is most common at the forest/grass-land margin because it nests in cavities of dead trees but feeds on insects and worms in open areas.

ASPEN

Is heat sensitive and easily killed by fire. However it sprouts vigorously and produces up to 60,000 new stems per acre after fire.

LODGEPOLE PINE

Some cones require heat to release their seeds. Where fire is frequent, more cones are of this type.

GOSHAWK

Lives within the coniferous forest and feeds mostly on birds. Builds its nest as a platform of sticks in a tree.

UINTA GROUND SQUIRREL

Lives in meadows and grasslands, feeding on grasses and weeds. Nests in underground burrows, hibernating in winter.

1   Woodruff Cooperative Wildlife Management Area
2   Rich County Bottoms
3   Round Valley
4   Bear Lake Overlook
5   Riverside Nature Trail
6   Rock Creek
7   Hardware Ranch
8   Porcupine Reservoir
9   Cutler Marsh
10 Wellsville Wilderness

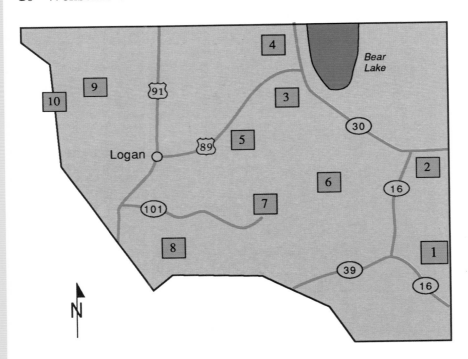

1 | Woodruff Cooperative Wildlife Management Area

Description: High desert site with sagebrush and grassland habitats. View wintering sage grouse, golden eagle, antelope, mule deer, and elk. This site is managed cooperatively by the Bureau of Land Management, the Utah Division of Wildlife Resources, and the Woodruff Livestock Association.

Viewing Information: Viewing probability is moderate from December through March. View wildlife from turnouts along highway. Binoculars required. Interpretive signs planned for 1990.

Directions: From Utah-Wyoming state line, follow Utah 16 north of Evanston, Wyoming. Viewing opportunity for the next four miles along east side of highway.

Ownership: BLM/UDWR (977-4300/479-5143)
Size: 4,000 acres **Closest Town:** Woodruff **P**

2 | Rich County Bottoms

Description: Large wetland, meadow, and farmland complex along the Bear River. Excellent viewing of ducks, geese, herons, egrets, sandhill cranes, swallows, harriers, and other birds.

Viewing Information: Viewing probability for featured species is high from April through October. Spring migration is best viewing, when birds are most numerous and in breeding plumage. Viewing site is entirely privately owned; roadside viewing only. HONOR THE RIGHTS OF PRIVATE LANDOWNERS AND VIEW ONLY FROM ROADSIDE TURNOUTS. PLEASE USE CAUTION WHEN PARKING ALONG ROADSIDES.

Directions: Loop route begins on Utah 16 about two miles north of Randolph. Travel north 7.5 miles to intersection of Utah 30 (Sage Creek Junction). Drive east 3.5 miles to Lincoln County, Wyoming Road 220, then travel south 6.4 miles on gravel surface to Crawford Mountain Road. Drive west five miles to point of origin on Utah 16.

Ownership: PVT (UDWR 479-5143)
Size: Twenty-two mile loop road **Closest Town:** Randolph

 The only Utah bird species which are predominantly blue in color are the Steller, scrub, and pinon jays, the western and mountain bluebirds, the blue grosbeak, and the indigo and lazuli buntings.

BRIDGERLAND

3 | **Round Valley**

Description: Picturesque marsh-farmland setting in secluded valley at the base of the east slope of the Bear River Range. Waterfowl, whistling swans, herons, sandhill cranes, egrets, golden eagles, red-tailed hawks, American kestrels, and occasionally bobolinks are among the bird species visible from mid-spring through fall. Moose, elk, mule deer, and red fox are most visible in early spring and fall.

Viewing Information: Viewing probability is high. Drive-by viewing only, as all lands are privately owned. HONOR THE RIGHTS OF PRIVATE LANDOWNERS AND VIEW ONLY FROM ROADSIDE TURNOUTS. PLEASE USE CAUTION WHEN PARKING ALONG ROADSIDES. The end of the tour is about one mile from Rendezvous Beach State Park.

Directions: Turn west off Utah 30 at Laketown, then turn west at Henry Early Park. Travel 1.2 miles to start of viewing loop. Turn south onto East Round Valley Drive, continue 4.6 miles to North Round Valley Drive. Follow for about five miles to Utah 30, completing loop tour.

Ownership: PVT (UDWR 479-5143)
Size: 4,200 acres **Closest Town:** Laketown

4 | **Bear Lake Overlook**

Description: Panoramic overlook of Bear Lake and its surroundings. Coniferous forest, mountain brush, and sagebrush habitats. Bear Lake itself offers viewing of unique bird species including loons and osprey during the spring and fall, and bald eagles during the late fall and winter. The nearby Limber Pine Trail offers viewing of warblers, finches, mountain bluebirds, western tanagers, blue grouse, and the unique purple martin. Also view mule deer and numerous small mammals including golden-mantled ground squirrels and red squirrels.

Viewing Information: Viewing probability is moderate from spring through fall. Visitor center to be constructed will feature interpretive information on the unique resident fish species of Bear Lake. Interpretive brochure available for Limber Pine Trail.

Directions: From Garden City, follow U.S. 89 west about five miles to the overlook on south side of highway. Limber Pine Trail, a one-mile hiking loop, is located about one mile west at Bear Lake Summit on south side of U.S. 89.

Ownership: USFS (753-2772)
Size: 160 acres   **Closest Town:** Garden City   **P**

The mourning dove is abundant in the deserts and dry uplands of Utah during the summer, particularly near water holes and streams. It usually migrates south by early September. J. KIRK GARDNER

BRIDGERLAND

5 | Riverside Nature Trail

Description: Foot-trail between the Spring Hollow and Malibu-Guinavah Campgrounds in the Wasatch-Cache National Forest. Trail passes through riparian vegetation adjacent to Logan River. Excellent viewing of belted kingfishers, violet-green swallows, house wrens, dippers, orange-crowned warblers, lazuli buntings, dark-eyed juncos, and many other bird species, as well as small mammals associated with conifer, aspen, and cottonwood habitats. Moose are occasional visitors.

Viewing Information: Viewing probability is high in spring and summer. Gentle hiking trail about one-and-one-half miles long. Brochure and species list available, developed in partnership with Bridgerland Audubon Society.

Directions: From Logan, follow U.S. 89 approximately six miles east and exit at Spring Hollow Campground. Hike Forest Trail 052 to Malibu-Guinavah Campground.

Ownership: USFS (753-2772)
Size: One-and-one-half mile trail **Closest Town:** Logan 　P⛱▲$🏠 🚶

6 | Rock Creek

Description: Gentle hike along riparian bottom into old-growth coniferous forest on adjacent north facing slopes. Aspen, mountain brush, and sagebrush on adjacent south facing slopes. Great wildlife diversity where these habitats converge. Common summer bird species include mallard, common snipe, white-crowned sparrow, and pine siskin. Summer mammals include red squirrel, beaver, chipmunk, moose, and mule deer.

Viewing Information: Viewing probability is high for birds and moderate for other species. Best bird viewing is June through July; June through September is best for other species. Note the wildlife species differences in the three major vegetative communities.

Directions: Turn west off of Utah 16 at Randolph and travel 2.2 miles to Little Creek Reservoir. Turn north onto Forest Road 058, then travel west for 12.5 miles to Forest Road 060. Turn south and travel 2.5 miles to Rock Creek turnoff, then drive another 2.7 miles south to parking area at Rock Creek. THIS IS A DIRT ROAD AND SHOULD BE AVOIDED WHEN WET.

Ownership: USFS (625-5112)
Size: Two-mile hike **Closest Town:** Randolph 　P

7 │ Hardware Ranch

Description: Elk wintering area on site of former livestock ranch managed by the Utah Division of Wildlife Resources. Primarily hay meadow and sagebrush habitats. Mule deer and moose use the general area during the winter, when bald eagles may also occasionally be viewed.

Viewing Information: Viewing probability is high. Winter viewing sleigh rides (fees charged) are provided to view 700 elk on feeding grounds. Call ahead to check on weather conditions and sleigh ride operation. Visitor center has interpretive materials and telescope.

Directions: At the junction of Utah 101 and Utah 165 (just east of Hyrum) turn east on Utah 101. Travel eighteen miles to the Hardware Ranch.

Ownership: UDWR (245-3131)
Size: 21,500 acres **Closest Town:** Hyrum   **P** 🚶

8 │ Porcupine Reservoir

Description: Kokanee spawning site in East Fork of Little Bear River above Porcupine Reservoir. Juniper and mountain brush setting on slopes adjacent to reservoir.

Viewing Information: High probability of viewing kokanee salmon in the stream habitat during September spawning runs. Fish are brightly colored and easily viewed. Fishing prohibited from August 16 through September 30. LANDS ADJACENT TO THE RESERVOIR ARE PRIVATE. PLEASE HONOR THE RIGHTS OF PRIVATE LANDOWNERS. DO NOT TRESPASS WITHOUT PERMISSION.

Directions: Follow Utah 165 south through Paradise and Avon. About 1 mile south of Avon, turn east and continue four miles to Porcupine Reservoir. Continue around reservoir to inlet (viewing site).

Ownership: PVT (UDWR 479-5143)
Size: Two acres **Closest Town:** Paradise   **P**

The last grizzly bear in Utah, a purported livestock killer known as "Old Ephraim," was killed in 1923 in the Logan River drainage.

| 9 | **Cutler Marsh** |

Description: Classic marsh habitat on Bear River in Cache Valley. Open water, bulrush, cattail, and meadow habitat supports a wide variety of waterfowl, wading birds, and shorebirds. Great site for viewing pelicans, herons, grebes, egrets, white-faced ibis, and sandhill cranes. The site is operated under license from the Federal Energy Regulatory Commission by Utah Power and Light Company and is available for public use.

Viewing Information: Viewing probability is high. Unique site for wildlife viewing from canoe or kayak. Also view from roadways along the Bear, Little Bear, and Logan Rivers; foot access is limited. Year-round viewing opportunity, but summer is best for viewing waterfowl in breeding plumage. Great blue herons, white-faced ibis, and snowy plovers nest around the marsh. ALL LANDS ARE PRIVATE OR LEASED. PLEASE DO NOT PARK IN FRONT OF GATES AND CLOSE THE GATES YOU OPEN. USE CAUTION WHEN PARKING ALONG ROADWAYS.

***Directions:** In Logan, turn west off Main Street onto Utah 30 (200 North). Travel about five miles to parking area and boat ramp at Bear River crossing. To reach Benson Marina, which has more parking and a boat ramp, drive north on 3000 West to 3000 North, then drive west to marina.*

Ownership: PVT (UDWR 479-5143)
Size: 4,800 acres **Closest Town:** Logan

Mink are found along rivers and streams in the mountains and valleys of northern and central Utah. They feed on a variety of small mammals, birds, and fish.
JAN L. WASSINK

| 10 | **Wellsville Wilderness** |

Description: Vista from the north end of the Wellsville Mountain Range with view north to Idaho border. View a wide variety of birds of prey including northern harriers, American kestrels, goshawks, Cooper's and sharp-shinned hawks, red-tailed hawks, golden eagles, and others in the fall.

Viewing Information: Viewing probability is high during late August through mid-October, with peak numbers observed in September. Binoculars are important to identify birds. Trail is steep (6% to 12%) and hike requires about two hours (one-way). Trail passes through conifers, aspen, and mountain brush to summit. NO DRINKING WATER AVAILABLE ON-SITE.

Directions: In Mendon, turn west off of Utah 23 at the corner of 300 North—100 West onto gravel surface road. Travel two miles to trailhead. LAST ONE-HALF MILE IS DIRT ROAD AND SHOULD BE AVOIDED WHEN WET. Hike Deep Canyon Trail to summit, then hike north one mile to vista.

Ownership: USFS (753-2772)
Size: Three-mile trail   **Closest Town:** Mendon   **P ⚐**

The American kestrel is one of the most commonly seen birds of prey in Utah. Frequently it can be identified by its characteristic hovering in search of prey in open country and farmland. J. KIRK GARDNER

11 **Clear Creek Campground**
12 **Salt Creek Waterfowl Management Area**
13 **Golden Spike National Historic Site**
14 **Willard Bay-Harold Crane Marsh**
15 **Ogden Bay Waterfowl Management Area**
16 **Ogden Nature Center**
17 **North Fork Park**
18 **Middle Fork Wildlife Management Area**
19 **North Arm**
20 **Beus Park**
21 **Davis Peaks**
22 **Morgan-Henefer Loop**

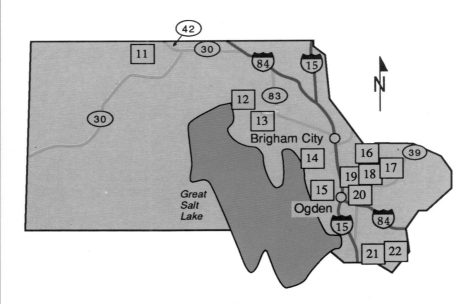

11 | Clear Creek Campground

Description: Cottonwood riparian habitat in pinyon-juniper woodland. Part of campground complex. Excellent opportunity to observe a variety of songbirds in northwestern mountains. Hawks, owls, squirrels, rabbits, hares, and deer may also be viewed.

Viewing Information: High probability of viewing riparian-dependent species in the arid surroundings during spring and early summer. Less species diversity and smaller populations in the fall.

Directions: *Leave Utah on Utah 42, entering Idaho on County Route 30-S. Drive about one mile and turn west onto gravel road at the old town site of Strevell, Idaho. Drive 3.3 miles and turn south at "Clear Creek Campground" sign. Travel another 5.5 miles to the Sawtooth National Forest boundary and begin Forest Road 001. Drive 0.7 miles to the campground entrance.*

Ownership: USFS (208 678-0430)
Size: 150 acres **Closest Town:** Snowville   P ⛺🏕🚶🏞⛰$

12 | Salt Creek Waterfowl Management Area

Description: Classic marsh habitat that provides nesting, brood rearing, and migration habitat for a wide variety of ducks, geese, cranes, herons, egrets, and various shorebirds.

Viewing Information: High probability of viewing wetland species from April through November; activity generally peaks in April and September. View entire management area from Compton's Knoll with spotting scope and/or binoculars.

Directions: *From Corinne, travel eight miles west on Utah 83 and turn north. Travel about 3.5 miles on gravel surface to the Salt Creek Waterfowl Management Area entrance. Continue 1.2 miles to Compton's Knoll.*

Ownership: UDWR (479-5143)
Size: 6,900 acres **Closest Town:** Corinne   P

 The northern oriole, a summer resident throughout Utah, builds a pouch-like nest, usually in tall, deciduous trees along streams and creeks.

13	**Golden Spike National Historic Site**

Description: Sagebrush and grasslands provide habitat for a wide variety of high desert species. Unique species include sharp-tailed grouse, sage grouse, and burrowing owl. More common species include northern harrier, American kestrel, red-tailed hawk, golden eagle, ring-necked pheasant, meadowlark, raven, jackrabbit, badger, and mule deer.

Viewing Information: This is the only sharp-tailed grouse viewing site listed in this guide; view in the visitor center vicinity during the March breeding season. View sage grouse on their strutting grounds from late March through mid-April. Spring is generally the best viewing period for most species, although viewing possible year-round. Viewing probability is moderate to limited for all species and varies with the season. Visitor center display features completion of the Pacific Railroad at Promontory Summit.

Directions: From Interstate 15 take exit 368 (Corinne), then travel west on Utah 83. Continue eighteen miles past Corinne, then turn west on paved county road and follow eight miles to visitor center. OR, from Interstate 84 take exit 26 and follow 83 south. Turn west at the Thiokol Plant, west again at county paved road across from the Thiokol Test Area, then drive eight miles to the visitor center.

Ownership: NPS (471-2209)
Size: 2,700 acres **Closest Town:** Corinne   P🏕️🅿️🚶🚻♿ $

Sharp-tailed grouse are limited to a few populations in northern Utah. However, they are commonly viewed at the Golden Spike National Monument during the spring breeding season. HARRY ENGELS

14 | Willard Bay-Harold Crane Marsh

Description: A major staging and migrating area for waterfowl and shorebirds. The area is part of the Bear River Bay of the Great Salt Lake, one of the world's largest wetlands. The area contains large blocks of natural mudflats important to shorebirds. Most of this habitat is visible from vehicle. Eagles, falcons, avocets, gulls, white-faced ibis, herons, egrets, plovers, sandpipers, and more. Upland bird habitat is managed for pheasants, quails, mourning doves, curlews, and a variety of other bird species.

Viewing Information: High probability of viewing wetland bird species from March through November, especially during migrational peaks. Best viewing is from roadway into the western portion of the area. Walk-in use also permitted. Bird list, maps, and informational brochure are planned.

Directions: From Interstate 15, take exit 354 (Willard Bay, Pleasant View). Turn west on 4000 North and drive to south marina of Willard Bay (walk-in access only). For vehicle access, drive to the end of 4000 North to gravel road. For another area access point, take Interstate 15 exit 347 and turn west onto Utah 39. Turn north on 6700 West to walk-in access. This access point is near the Ogden Bay Waterfowl Management Area north-end access points.

Ownership: UDWR (479-5143)
Size: 12,500 acres **Closest Town:** Plain City **P** 🚶 ⛵

The beautifully marked American avocet is abundant in the marsh and mudflat habitat of the state. It lays three to four olive, blotched eggs each year. JAN L. WASSINK

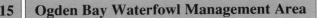

| 15 | Ogden Bay Waterfowl Management Area |

Description: One of the most productive wetland areas in the nation. Habitat to a wide variety of grebes, herons, ducks, geese, raptors, plovers, sandpipers, gulls, and terns, as well as other bird and mammalian species. Much wetland habitat now in the recovery process after the rising Great Salt Lake flooded the marshes with saline water in 1983. Some 15,000 ducks are produced annually here; common nesting species include redhead, cinnamon teal, mallard, gadwall, pintail and northern shoveler.

Viewing Information: High probability of viewing wetland species from March through November. Migrating duck numbers usually peak in September, occasionally reaching one-half million birds. A vehicle loop should be open to viewers by late summer, 1991. Hikers may walk dike roads past vehicle gates year-round. Bird lists and other interpretive materials are available; guided tours are available to large groups.

Directions: To access area headquarters, take exit 341 (Roy) off Interstate 15 and turn west on Utah 97. Drive to the end Utah 97, turn north on Utah 108 for 0.1 mile, then west on Utah 98 to Hooper. Continue west on 5500 South, then north on 7500 West. Other south-end access points are located on 5500 West and 5100 West. To access the north end, take Interstate 15 exit 347, turn west on Utah 39, then south on 7500 West or 9500 West.

Ownership: UDWR (479-5143)
Size: 20,000 acres **Closest Town:** Hooper   P 🏕 🚶 🏕

Several pintail drakes bank around a single hen during an aerial courtship display. Pintails are one of several nesting species at Odgen Bay Waterfowl Management Area, one of Utah's most productive wetland sites. LARRY R. DITTO

Description: A privately operated nature reserve that features a variety of wildlife habitats including wetlands, riparian, and shrub/grasslands in an urban setting. Bird species include great blue heron, snowy egret, willet, Canada geese, wood duck, red-winged blackbird, and red-tailed hawk. Mammalian species include raccoon, muskrat, red fox, porcupine, and mule deer. A trail system takes visitors through the various habitats for year-round viewing. The site is operated by the Ogden Nature Center.

Viewing Information: High probability of viewing most species in the appropriate season. Greatest species diversity during the spring and early summer, although fall migration increases species numbers also. Center features a naturalist on duty, interpretive center, natural history museum, wildlife rehabilitation center, and guided nature tours. Membership or entrance fee required.

Directions: From Interstate 15 take exit 347 at Ogden (12th Street). Drive approximately 1/2 mile east and turn into the Nature Center at 966 12th Street.

Ownership: NP (621-7595)
Size: 127 acres **Closest Town:** Ogden

P ⚐ 🚶 ⛱ $

The most common fur-bearing species in Utah, the muskrat can be found in marsh habitat throughout the state. Its flat tail (when viewed from side to side) is unique among mammals. JAN L. WASSINK

Prolonged winter conditions including deep snow and cold temperatures make winter survival difficult for mule deer, particularly when winter forage or body fat reserves are depleted. The Middle Fork Wildlife Management Area is one of several locations in Utah to view wintering mule deer and other animals.
CHRISTOPHER CAUBLE

17 | North Fork Park

Description: Scenic area on east side of Wasatch Mountains at the north end of Ogden Valley. Park is managed by Weber County Parks Department. One of the best examples of mature mountain brush species including maple, oakbrush, chokecherry, and aspen. Also conifer and some cliff habitats. FALL COLORS ARE EXTRAORDINARY HERE. A diversity of wildlife species including mule deer, moose, golden eagle, red-tailed hawk, ruffed and blue grouse, and many songbirds may be observed here.

Viewing Information: Viewing probability is moderate to limited for most species. Best viewing for species diversity is June and July, but viewing into the fall is also good. Park serves as a trailhead into the high, mountainous areas (e.g. Ben Lomond Peak) of the Wasatch-Cache National Forest. Park trails are accessible to hikers and horses. Park is open from mid-May through mid-October and has group fee areas.

Directions: Continue north at the north end of Utah 162 (3300 East 4100 North), which is northwest of Liberty, for 1.4 miles. Then take the northwest fork at the North Fork/Avon "Y" junction. Travel another 1.1 miles and turn northwest at the sign for North Fork Park. Continue another one mile to the Park entrance.

Ownership: NP (399-8491)
Size: 2,400 acres **Closest town:** Eden

18 | Middle Fork Wildlife Management Area

Description: Winter range for elk, mule deer, and moose. Sagebrush, mountain brush, oakbrush, grassland, and riparian vegetative communities.

Viewing Information: Moderate probability of viewing target species from December through mid-April. View from entrance gate during the winter, or hike a short distance into the area during the spring. Wintering bald eagles may occasionally be viewed from this vantage point. Year-round public access permitted to hikers and horseback riders on the established trail system only. MOTORIZED VEHICLES PROHIBITED.

Directions: Just east of Huntsville, where Utah 39 turns east to the South Fork of the Ogden River, drive north about 0.3 miles on Utah 166. Then turn onto 7800 East (paved county road) as Utah 166 turns to the west. Follow this county road two miles to the entrance (signed) of the wildlife management area.

Ownership: UDWR (479-5143)
Size: 10,000 acres **Closest Town:** Huntsville

19 | North Arm

Description: Trail system through riparian-wetland area featuring dense cottonwood/brushland and marsh vegetative communities. The site is the inlet of the North Fork Ogden River into Pineview Reservoir and is managed by the Wasatch-Cache National Forest. A wide variety of songbirds including yellow warbler, Wilson's warbler, lazuli bunting, white-crowned sparrow, and northern oriole are common summer residents. Waterfowl species observed here include Canada geese and a variety of ducks; raptors include bald eagle, red-tailed hawk, Cooper's and sharp-shinned hawk, goshawk, and American kestrel. Mule deer and moose occasionally use the site as well.

Viewing Information: High probability of viewing songbirds and wetland bird species during the spring and early summer. Raptor and small mammal viewing probability is moderate and large mammal viewing probability is limited during the summer. Interpretive signs, brochure, and viewing trail will be available in late summer, 1990. The extent of marsh habitat varies with Pineview Reservoir water level.

Directions: Turn north off Utah 39 to Utah 162 at the Pineview Reservoir Dam. Drive 3.8 miles to the North Arm parking area just across the North Fork of the Ogden River.

Ownership: USFS (625-5112)
Size: 100 acres **Closest Town:** Huntsville **P** ☂

20 | Beus Park

Description: Natural wetland area in urban setting managed by Ogden City Parks Department. Cottonwood and oakbrush vegetation types skirt the pond where bulrush and cattails are the dominant vegetation. A good stopover for migrating birds, the area has had 85 identified birds species. The pond also hosts bass, bluegill, black bullhead, and gambusia.

Viewing Information: Site offers year-round viewing opportunity. High probability of viewing bird species all year, but best during spring, early summer, and fall. Viewing probability is limited for fish species. Paved trail traverses the area. Nature trail brochure and bird list is available.

Directions: In Ogden, drive east on 42nd Street from Harrison Boulevard (Utah 203). The roadway curves to the north at Beus Park. Look for the entrance sign and parking lot on the right after making the curve.

Ownership: NP (629-8284)
Size: Twelve acres **Closest Town:** Ogden   P⛺🚻☂⛅♿

21 | Davis Peaks

Description: Spectacular ridgetop drive from Bountiful to Farmington overlooking the Great Salt Lake. Excellent on-site scenery. Mountain brush, aspen, coniferous forest, and grassland vegetative communities. FALL COLORS ARE EXTRAORDINARY HERE. Wildlife species include mule deer, blue and ruffed grouse, red-tailed hawk, golden eagle, and American kestrel. Also see snowshoe hares, porcupines, bobcats, coyotes, and a wide variety of songbirds.

Viewing Information: High probability of viewing deer, grouse, raptors, and songbirds from mid-June through mid-November. Viewing probability for other species is limited. Use turnouts and roadside pull-offs for maximum viewing. Traffic may be relatively heavy at times and ROAD IS NARROW IN PLACES (e.g. Farmington Canyon). OBSERVE ALL TRAFFIC LAWS AND DRIVE CAUTIOUSLY. Binoculars are essential.

Directions: Exit Interstate 15 at 400 North in Bountiful, then travel east to 1300 East. Turn north to 600 East, then east onto Skyline Drive and take the switchbacks that are below the white "B" on the hillside, and follow gravel road east (begin loop drive on Forest Road 177) to ridgetop. Continue north 12.5 miles from end of pavement to Bountiful Peak overlook. Continue north, then west down Farmington Canyon (another eleven miles). End viewing loop at 600 North and 100 East in Farmington. WITH FREQUENT STOPS, THIS IS A THREE-HOUR TRIP ON A ROUGH, GRAVEL SURFACE ROAD. THE ROAD IS CLOSED BY SNOW FROM DECEMBER THROUGH MAY.

Ownership: USFS (524-5042)
Size: Twenty-five mile drive **Closest Town:** Bountiful or Farmington **P𝑘▲**

Color changes in Gambel oak, bigtooth maple, and aspen mark the beginning of autumn along the Wasatch Front. Scenes like this can be viewed at the Davis Peaks and North Fork Park viewing sites in late September.
BRUCE ANDERSEN

22 | Morgan-Henefer Loop

Description: Loop drive through riparian/farmland area along East Canyon Creek to East Canyon State Park and reservoir, then northeast through sagebrush and grassland vegetation to Henefer. Bald eagles winter in the vicinity of East Canyon reservoir and along the Weber River. Mule deer winter in East Canyon and downstream from the reservoir. View sage grouse from the road near the Morgan-Summit County line. Kokanee salmon spawn in the inlet to East Canyon Reservoir in September.

Viewing Information: High to moderate probability of viewing featured species in appropriate seasons. An array of songbirds and waterfowl may also be observed during the early summer at the reservoir. Jackrabbits, snowshoe hares, and cottontail rabbits may also be viewed along this route. Although a year-round viewing opportunity, WINTER SNOW CONDITIONS MAY PREVENT TRAVEL ON THE ENTIRE LOOP. Much of the route is private land, therefore HONOR THE RIGHTS OF PRIVATE LANDOWNERS AND ASK PERMISSION FOR ENTRY.

Directions: Exit Interstate 84 at Morgan and take Utah 66 to East Canyon Reservoir. Turn south on Utah 65 to the south end of the reservoir. Take Utah 65 north to Henefer and return to Morgan via Interstate 84.

Ownership: PVT/UDPR (UDPR 829-6866)
Size: Thirty-two mile drive **Closest Town:** Morgan

The color of the whitetail jackrabbit turns from a deep brown in summer to snow white in winter. It is occasionally confused with its close relative, the snowshoe hare, which is smaller and has shorter ears. CHRISTOPHER CAUBLE

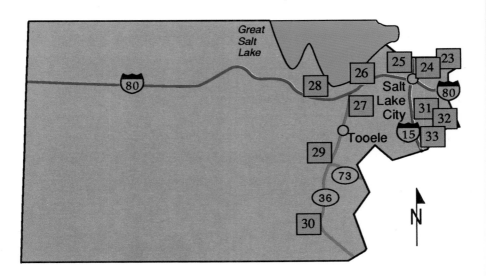

23 | Pioneer Trail State Park

Description: Urban historic state park located in foothills of Wasatch Mountains. Vegetative plant communities include oakbrush and grassland types. Habitat for diverse wildlife species including cottontail rabbit, ground squirrel, red fox, mule deer, raccoon, scrub jay, green-tailed towhee, Cooper's hawk, and red-tailed hawk.

Viewing Information: Moderate probability of viewing mule deer and resident bird species during winter and spring, and for small mammals, songbirds, and raptors during spring through fall. Primary viewing is north of the paved service road in and around the Old Deseret "settlement."

Directions: In Salt Lake City, turn off Utah 186 (Foothill Boulevard) east to Sunnyside Avenue. Drive east a short distance to the entrance to the Park at 2601 East Sunnyside Avenue.

Ownership: UDPR (584-8392)
Size: 450 acres **Closest Town:** Salt Lake City

P 🏕 ⛺ $

Badgers are typically found in open grasslands and brushy areas, where they use their powerful legs and long front claws to dig out ground squirrels and other rodents. Early morning is the best time to look for badgers, which are mostly active at night. JEFF FOOTT

24 | Hotel Utah

Description: Downtown Salt Lake City view of nesting peregrine falcon pair. Peregrine nestboxes located on south side and northeast corner of Hotel Utah. Nestboxes will be removed for hotel renovation during 1991 and 1992, when they will be placed atop one to three tall office buildings along South Temple, State, or Main Streets. Peregrines may also be viewed on LDS Church Office Building, Crossroads Plaza Tower, Beneficial Life Tower, and Eagle Gate Tower. Other species present include swallows and nighthawks.

Viewing Information: Viewing probability is high during the period when young are hatched to fledging; otherwise, viewing is moderate to limited. The birds arrive in late March to early April. Eggs are hatched from mid to late May and birds are fledged from mid to late June. A peregrine falcon watchpost is maintained by Utah Division of Wildlife Resources personnel and volunteers immediately south of the Hotel from the egg-laying period to after the young are fledged. Watchpost has binoculars and spotting scopes, live television of nest box activities, and informational brochures.

Directions: *From the corner of Main Street and South Temple in Salt Lake City, walk east on the south side of the street. Continue east until directly opposite the Hotel Utah building. Look for peregrine falcon watchpost.*

Ownership: PVT (UDWR 596-8660)
Closest Town: Salt Lake City   P 🏨 🍴 🛏 ♿

Window ledges on tall office buildings may closely resemble typical natural nesting habitat (ledges on tall cliffs) for the peregrine falcon. The famous downtown Salt Lake City pair are shown here on the Eagle Gate Tower. MARGUERITE ROBERTS

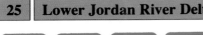

25 | **Lower Jordan River Delta**

Description: Two sites, Jordan River Parkway and Salt Lake City Water Reclamation Plant, with year-round viewing in urban riparian/wetland complex. Great variety of shorebirds and wading birds, especially during the spring. In winter, view bald eagles and waterfowl; in summer, view beaver, muskrat, and breeding wetland birds. Riparian corridor of willows, cottonwoods, and maples is excellent for migrating and nesting songbirds.

Viewing Information: High probability of viewing all species and groups during appropriate seasons. Jordan River delta, one of three freshwater flows into the Great Salt Lake, forms habitat base. Parts of Jordan River can be accessed by canoe for viewing.

Directions: For Jordan River Parkway site, take exit 25 (Salt Lake City, 22nd North) off Interstate 215, travel north about one mile on Rose Park Lane to Utah Division of Parks and Recreation building at north end of road. For Water Reclamation Plant site, turn east off of Redwood Road in Salt Lake City onto 2300 North. Site is on south side of road about one-half mile from turnoff.

Ownership: UDPR (533-4496)
Size: 180 acres **Closest Town:** Salt Lake City P 🏠 🚶 🏕

26 | **Great Salt Lake Shore**

Description: Southeastern shoreline of the Great Salt Lake. Year-round viewing of wetland bird species; target species change with seasons and water levels. Great Salt Lake has international significance to shorebird populations.

Viewing Information: High probability of viewing all species in appropriate season. Winter features northern shoveler and numerous diving ducks, gulls, and bald eagles; spring and summer host vast numbers of shorebirds including snowy plovers, American avocets, black-necked stilts, Wilson's phalaropes, gulls, terns, and waterfowl; fall migration features waterfowl. Easy viewing from frontage road adjacent to Interstate 80, or from the Great Salt Lake State Park beach.

Directions: Take Interstate 80 exit 104 to Saltair Resort. Turn north on beach frontage road and drive about one-half mile for viewing where canal crosses under roadway. Continue northeast along frontage road to Great Salt Lake State Park (fee area). In springtime, continue east on frontage road to the Salt Lake International Center and view foraging shorebirds and wading birds along north side of road.

Ownership: UDPR (533-4080)
Size: Four-mile drive **Closest Town:** Salt Lake City P 🏕

27 | Lake Point

Description: Elk winter range in highly visible setting at the northwest corner of the Oquirrh Mountains. Mountain brush vegetation is winter habitat for elk.

Viewing Information: Moderate probability of viewing elk from December through early March. Binoculars are essential for best viewing. ALL LANDS ARE PRIVATE. PLEASE HONOR THE RIGHTS OF PRIVATE LANDOWNERS AND DO NOT TRESPASS.

Directions: Exit Interstate 80 at the Tooele (Utah 36) Exit. Take the first road east to Lake Point and view groups of elk on the west slopes of the Oquirrhs.

Ownership: PVT (596-8660)
Size: 8,000 acres **Closest Town:** Tooele

The white-faced ibis is a common sight in the mudflats and farmlands along the Wasatch Front in summer. Their flight in long, wavering lines to and from their roost is characteristic of the species. JAN L. WASSINK

28 | Timpie Springs

Description: A spring-fed waterfowl management area on the south shore of the Great Salt Lake. Unique wetland situated between a highly saline portion of the Great Salt Lake and the west desert. The abundance of shorebirds, waterfowl, and wading birds varies with season of year. Peregrine falcon hack tower (species restoration site) is located on-site and is now used for nesting.

Viewing Information: High probability of viewing wetland bird species during spring through fall. Limited to moderate probability of viewing peregrine falcon from April through September. Trail system (dikes) accessible for public viewing from October through December; contact Utah Division of Wildlife Resources for entry information during other periods. The spring head, managed by the Salt Lake District, Bureau of Land Management, is located south of Interstate 80 and is open to viewing at all times.

Directions: From Salt Lake City, take Interstate 80 west and exit at Rowley. Turn north, proceed about one mile, and follow signs to viewing site. Park near locked gate entrance.

Ownership: UDWR (596-8660)
Size: 1,440 acres **Closest town:** Grantsville   **P** ⚐

29 | Rush Lake

Description: Ephemeral desert lake with wide variety of wetland bird species including common loon, double-crested cormorant, heron, egret, and western, eared, and pied-billed grebes.

Viewing Information: High probability of viewing wetland bird species from spring to fall. Best viewing of greatest number of species in the spring. Shorebird populations tend to increase as the lake level decreases. SOME ADJACENT LANDS ARE PRIVATE. PLEASE HONOR THE RIGHTS OF PRIVATE LANDOWNERS AND DO NOT TRESPASS.

Directions: From Tooele, take Utah 36 south to Stockton, then turn west. View from gravel road along the north and west sides of lake or continue south on Utah 36 and view wildlife from highway turnouts.

Ownership: PVT/BLM (BLM 977-4300)
Size: 2,000 acres **Closest Town:** Stockton   **P**

 Over 1,200 bald eagles winter in Utah each year. Normally fish-eaters, the Rush Valley population is unique in that the eagles survive the winters by feeding on rabbits and deer carrion.

30 | Vernon

Description: Winter viewing of various raptors including bald eagle, golden eagle, and rough-legged hawk in a sagebrush/farmland area. Birds roost in the relatively few cottonwood trees in otherwise desert habitat. Raptors are drawn to the area by large blacktail jackrabbit populations. Mule deer and antelope may occasionally be observed.

Viewing Information: High probability of viewing raptors from late December to mid-March. View birds from paved and gravel roads south and west of Vernon; roadside parking only.

Directions: From Tooele, take Utah 36 south to Vernon. Take county roads south and west. SOME LANDS ARE PRIVATE. PLEASE HONOR THE RIGHTS OF PRIVATE LANDOWNERS AND DO NOT TRESPASS.

Ownership: BLM (977-4300)
Size: 5,000 acres **Closest Town:** Vernon

31 | Dimple Dell Regional Park

Description: Foot and horseback trail system through natural area in urban setting managed by the Salt Lake County Parks and Recreation Department. Oakbrush, bitterbrush, grassland, and riparian vegetative communities. In winter, view mule deer, red fox, and many resident bird species. During spring and early summer, view bird species including green-tailed and rufous-sided towhee, common nighthawk, yellow-breasted chat, blue-gray gnatcatcher, sharp-shinned hawk, and broad-tailed hummingbird.

Viewing Information: Moderate probability of viewing all species in season. A variety of amphibians and reptiles including six lizard species may also be observed here during spring and summer.

Directions: From Interstate 15, take exit 298 at 9000 South. Drive east on Utah 209, then turn south on 3000 East and follow to the northeast park entrance at 9400 South (this is the preferred entrance). OR take Interstate 15 exit 297 at 10600 South. Drive east to 1300 East, then north to approximately 10500 South and enter the Wrangler Trailhead (the park's northwest entrance). OR, use the park's southeast entrance at approximately 2700 Dimple Dell Road.

Ownership: NP (468-2299)
Size: 643 acres **Closest Town:** Sandy P⊼𝍫

32 | Snowbird Mountain Trail

Description: Short hiking trail located in scenic Little Cottonwood Canyon at the Snowbird Ski Resort. The site offers a unique glimpse of the Salt Lake Valley below and is surrounded by alpine peaks. Trail is located adjacent to riparian zone of Little Cottonwood Creek. Mixed conifer-aspen and grassland form the habitat base. Wide variety of bird species including hummingbirds, warblers, kinglets, vireos, woodpeckers, American kestrels, Cooper's hawks, goshawks, red-tailed hawks, and golden eagles. Mammals here include red squirrels, Uinta ground squirrels, chipmunks, an occasional weasel, and mule deer.

Viewing Information: Moderate to limited probability of viewing all species from late May through October. Interpretive center is located in the ticket sales area of the Tram Building at Snowbird during the viewing period. The trail grade is designed to accommodate the physically challenged and includes various interpretive aids. The trail will be completely equipped and interpreted by mid-summer, 1990.

Directions: From Salt Lake City, drive east on Utah 210 to the Snowbird Resort. Begin tour at Tram Building.

Ownership: USFS (524-5042)
Size: 1-1/2 mile trail **Closest Town:** Salt Lake City   P 🏠 🚻 🛏 ♿ 🚶

33 | White Pine Lake

Description: Hike to scenic viewing site at 10,000-foot elevation lake in the Wasatch-Cache National Forest. Alpine-fir habitat adjacent to talus slopes (rock slides). Pika is the featured species here. Other small mammals include snowshoe hare, Uinta ground squirrel, golden-mantled ground squirrel, and marmot. Pine grosbeak, Clark's nutcracker, red crossbill, black rosy finch, and a variety of warblers, hummingbirds, and flycatchers may also be viewed here.

Viewing Information: Moderate probability of viewing pika during summer and early fall in its primary habitat, the talus slopes. The pika, a cottontail rabbit-sized animal, may be identified by their rounded ears and no visible tail. Note their habit of cutting and piling grasses and forbs (hay) to store as a winter food source.

Directions: Follow Utah 210 east of Salt Lake to Little Cottonwood Canyon. The trailhead for White Pine Lake is about .75 miles east of Tanners Flat Campground. HIKE THIS TRAIL TO WHITE PINE LAKE (approximately three miles). Good pika habitat in the talus slopes around the west and south sides of the lake.

Ownership: USFS (524-5042)
Size: 200 acres **Closest Town:** Salt Lake City   P 🏠 🚶

34 **Mount Timpanogos Wilderness**
35 **Cascade Springs**
36 **Bridal Veil Falls**
37 **Provo Bay**
38 **Steele Ranch**
39 **Mount Nebo Scenic Loop**
40 **Indianola Wildlife Management Area**
41 **Strawberry Valley**
42 **Rockport State Park**
43 **Henefer-Echo Wildlife Management Area**
44 **Hole-in-the-Rock**
45 **Ptarmigan Loop**
46 **Whitney Basin**
47 **Bald Mountain**

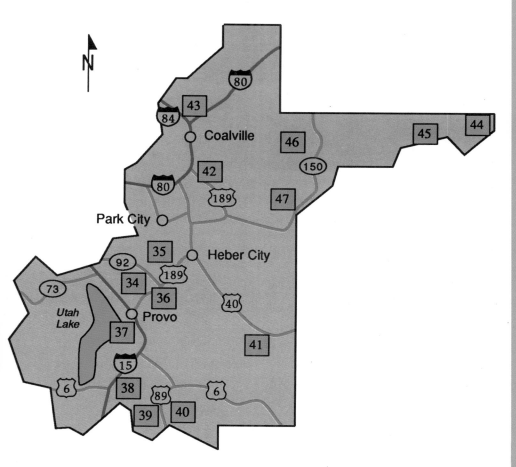

34 | Mount Timpanogos Wilderness

Description: Wilderness hike through beautiful, rugged alpine landscape. Rock-cliff, conifer, aspen, mountain brush, and riparian habitats. Featured species include mountain goat, mule deer, red squirrel, marmot, snowshoe hare, and a variety of songbirds and raptors.

Viewing Information: Probability of viewing featured species ranges from high to limited, depending upon viewing season and habitat type. Hikes range from easy to difficult. TRAILS MAY BE STEEP AND ARE AT HIGH ELEVATIONS; HIKERS SHOULD BE PREPARED.

Directions: Two trailheads access the Wilderness. The Timpooneke trailhead takes horseback riders and hikers to Emerald Lake (6.5 miles). Reach this trailhead by taking Exit 287 from Interstate 15 and turn east on Utah 92. Travel 16 miles to Timpooneke Campground turnoff. The trailhead is located .25 mile down the campground road. The Aspen Grove trailhead allows only hikers on the trail to Emerald Lake (5.3 miles). To reach this trailhead take Exit 275 (Orem 8th North) from Interstate 15 and drive east 3.5 miles to Utah 189. Continue east through Provo Canyon seven miles to the Sundance turnoff (Utah 92). Drive northeast six miles to the trailhead.

Ownership: USFS (785-3563)
Size: 10,750 acres **Closest Town:** Pleasant Grove or Orem P △ ⚏ ▲ ⚲

35 | Cascade Springs

Description: One-of-a-kind viewing site in the Uinta National Forest. Highly developed trail system in a beautiful riparian setting. A number of springs surface and flow into the Provo River, discharging seven million gallons of water daily. Rainbow trout visible in pools from boardwalks and bridges. A wide variety of songbirds, as well as frogs, toads, squirrels, and chipmunks are present.

Viewing Information: High probability of viewing trout in season. Also, high probability of viewing breeding birds in the spring and early summer. Viewing season runs from June through October; June and July are best. Interpretive trail with brochure on-site.

***Directions:** Take Utah 92 (Alpine Scenic Loop) and turn off onto Forest Road 114 to Cascade Springs. Drive about eight miles to the site.*

Ownership: USFS (785-3563)
Size: Twenty-five acres **Closest Town:** Midway P △ ♿ ⚲

36 | Bridal Veil Falls

Description: Tram ride (fee charged) to view Rocky Mountain goats across Provo River. Habitat is typically alpine vegetation scattered in rugged, rocky canyon. Spectacular view during spring and fall. Viewing is long distance; binoculars or spotting scope are required.

Viewing Information: High probability of viewing goats from tram opening in spring through mid-June and again in October until tram closing in late October. Mountain goats may occasionally be viewed from parking area at base of tram. Binoculars are available for rental or purchase.

Directions: From Orem, follow U.S. 189 east for about five miles to the site.

Ownership: PVT (225-4461)
Closest Town: Orem

P🏠🚻🅿♿$

The mountain goat was reintroduced to Lone Peak in the Wasatch Range in 1967 Additional populations now are established on Mount Timpanogos, the Uinta Mountains, and the Tushar Mountains. MICHAEL S. SAMPLE

37 Provo Bay

Description: Marsh area adjacent to Utah Lake. Wide array of wetland bird species may be observed from March through November including white pelican, snowy and cattle egrets, great blue heron, common tern, white-faced ibis, plus a variety of ducks, geese, and swallows.

Viewing Information: High viewing probability spring through fall; April and November migration peaks are best. HIKE ALONG DIKES FROM VARIOUS PARKING LOCATIONS.

Directions: Take Interstate 15 exit 263 and turn west onto Utah 77. Travel west to the Spanish Fork River bridge, then take dirt road north to Provo Bay. Park vehicle and walk into marsh on numerous trails. THERE IS A SIGNIFICANT AMOUNT OF PRIVATE LAND IN THE AREA. USE THIS ACCESS POINT ONLY OR SEEK PERMISSION TO ACCESS OTHER AREAS ON THE BAY.

Ownership: PVT/UDWR (UDWR 489-5678)
Size: 6,000 acres **Closest Town:** Provo   **P** 🚶

38 Steele Ranch

Description: A sagebrush and oakbrush winter range for elk and mule deer with easy access from the Interstate. Between 50 and 150 elk and numerous mule deer may be viewed here.

Viewing Information: High to moderate probability of viewing elk and mule deer from December through March. Use binoculars. VIEW ANIMALS FROM VEHICLE SO AS NOT TO DISTURB THEM.

Directions: From Santaquin, take Interstate 15 south to the first exit (about two miles). Drive south on frontage road on west side of Interstate 15 for about three miles. Turn east at Utah Division of Wildlife Resources sign, go under the Interstate, and drive into the graveled parking area.

Ownership: UDWR (489-5678)
Size: 1,000 acres **Closest Town:** Santaquin

Some 37 species of swans, geese, and ducks have been observed in Utah, which is the second driest state.

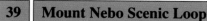

| 39 | **Mount Nebo Scenic Loop** |

Description: Spectacular mountain drive from juniper woodland through alpine habitat. A wide variety of wildlife viewing opportunities at turnouts and at numerous area trails. Viewing season is May through October. Bird species, including scrub and Steller's jay, western tanager, dark-eyed junco, white crowned sparrow, and ruby-crowned kinglet are most numerous in the spring and early summer. Raptors are common throughout the viewing season, but usually most numerous during migration in September and early October. Mule deer, elk, and sometimes moose may also be viewed in the spring and early summer.

Viewing Information: Viewing probability is high where visitors stop at turnouts and make short hikes from one of the fifteen trailheads. Use binoculars for best viewing. Interpretive materials on the scenic drive are available from Uinta National Forest offices.

Directions: From Nephi, follow Utah 132 east for about five miles and turn north at the Scenic Loop sign. OR, access the north end of the Loop from Utah 91 in Payson, following the signs.

Ownership: USFS (798-3571)
Size: Thirty-five mile drive **Closest Town:** Payson or Nephi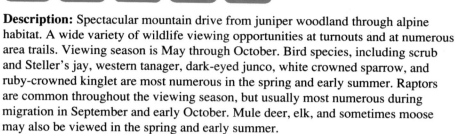

| 40 | **Indianola Wildlife Management Area** |

Description: Juniper woodland and sagebrush habitats that provide winter habitat for approximately 200 elk and many mule deer.

Viewing Information: Moderate probability of viewing elk and high probability of viewing mule deer from December through March. Roadside viewing opportunity; use binoculars for best view. USE CAUTION WHEN PARKING OR PULLING ONTO ROADWAY.

Directions: From its intersection with U.S. 6, travel south on U.S. 89 for approximately 15 miles to Sanpete County line. View animals just north of county line.

Ownership: UDWR (489-5678)
Size: 2,200 acres **Closest Town:** Fairview

The red crossbill, most abundant in Utah in the Uinta Mountains, uses its crossed mandibles to extract seeds from the cones of evergreen trees.

41 | Strawberry Valley

Description: Drive-by viewing in beautiful section of the Uinta National Forest, with great wildlife diversity. Spawning cutthroat trout may be viewed on any west-side tributary to Strawberry Reservoir during May and June; Crooked Creek and Streeper Creek are best. Sandhill cranes may be viewed in May and June near the Chicken Creek West and Chipman Creek areas. Waterfowl, herons, and egrets may be seen from spring through fall around the bay near the visitor center. Great viewing of elk and mule deer anywhere in the valley along the aspen edges in spring and early summer. Sage grouse may be seen in the Stinking Springs/Windy Ridge area from April through June. Red-tailed hawk, Swainson's hawk, and goshawk are commonly seen throughout the spring and summer. Uinta ground squirrels are common during the spring and early summer and serve as a prey base for many predators and raptors. Other species include dipper, white pelican, double-crested cormorant, spotted sandpiper, badger, chipmunk, and more.

Viewing Information: High to moderate probability of viewing the above-noted species in appropriate seasons. Visitor center located just off U.S. 40 on the the reservoir's west side. Utah Division of Wildlife Resources trout egg-taking station located near the visitor center.

Directions: From Heber City, drive twenty-three miles east on U.S. 40. Turn off on Forest Road 131 near visitor center. Chicken Creek West area is an additional 1.5 miles east on U.S. 40, then south another 1.5 miles to parking area.

Ownership: USFS (654-0470)
Size: 60,000 acres
Closest Town: Heber City

The fall colors of aspen abound throughout the mountains of Utah. Some of the best aspen habitat may be viewed in Strawberry Valley. TIM CLARK

42 | Rockport State Park

Description: State park in juniper woodland at Rockport Reservoir. Great diversity of bird species at inlet where mud flats, marsh, and riparian habitats occur. Great blue heron, western grebe, common loon, killdeer, snowy egret, common tern, great-horned owl, red-tailed hawk, broad-tailed hummingbird, yellow warbler, western kingbird, and dipper are common in the spring and early summer. November migrants include whistling swan and Canada geese. Uinta ground squirrel, least chipmunk, yellow-bellied marmot, badger, whitetail jackrabbit, and cottontail rabbit are commonly observed in summer. During winter, view bald eagles and mule deer.

Viewing Information: Probability of viewing species discussed above ranges from high to moderate in the appropriate season. Winter ski-trail offers unique opportunity to view mule deer and bald eagles on-site. Wildlife species list is available.

Directions: Exit Interstate 80 at Wanship, then drive south on U.S. 189 about six miles to the park entrance.

Ownership: UDPR (336-2241)
Size: 1,850 acres **Closest Town:** Wanship

P 🏠 ▲ 🏕 🛥 $ ♿

43 | Henefer-Echo Wildlife Management Area

Description: Wildlife management area administered primarily as winter range for elk, moose, and mule deer. Many small mammals, songbirds, upland birds, and raptors also use the site. Bald eagles are featured during the fall and winter. The habitat base is primarily mountain brush, sagebrush, and grassland, with some aspen.

Viewing Information: High to moderate probability of viewing elk, deer, and bald eagles during appropriate seasons. THE AREA IS CLOSED TO MOTORIZED VEHICLES. Only hiking, horseback riding, and cross-country skiing are permitted. Three separate viewing sites are described below.

Directions: From Interstate 84, take exit 115 (Henefer). Turn east under highway. For SITE ONE (spring-fall site), drive about three miles, turn north just south of the Croyden Cemetery, then follow gravel surface road 0.8 miles to the trailhead to Harris Canyon. For SITE TWO (spring-fall site), drive about one mile and turn south to the Fire Canyon trailhead. For SITE THREE (winter-spring site), continue south approximately one more mile to the Witch Rocks site. The Witch Rocks themselves are PRIVATE PROPERTY AND PERMISSION SHOULD BE OBTAINED PRIOR TO ENTRY. This is a roadside viewing opportunity.

Ownership: UDWR (479-5143)
Size: 14,000 acres **Closest Town:** Henefer P 🚶

44 Hole-in-the-Rock

Description: Roadside viewing of bighorn sheep, moose, mule deer, and occasionally elk. Sagebrush/grassland, coniferous forest, riparian willow, and cliff-face habitats. Bighorn sheep have recently been reintroduced here.

Viewing Information: Moderate to limited probability of viewing moose and bighorn sheep from June through October. Binoculars are needed for best viewing. Sagebrush/grassland slopes adjacent to south-facing cliffs are classic bighorn habitat.

Directions: From Mountain View, Wyoming, travel east on Wyoming 414 to two miles east of Lone Tree. Turn south on Uinta County 295, then drive seven miles to the Hoop Lake turnoff and follow to Forest Road 058. From this point turn south and view along Hole-in-the-Rock Mountain on drive to Hoop Lake OR continue west two miles and view along Beaver Mountain. THE ROAD FROM LONE TREE TO THE HOOP LAKE TURNOFF IS A DIRT ROAD AND IS VERY SLIPPERY WHEN WET. AVOID USING THIS ROAD WHEN WET!

Ownership: USFS (307 782-6555)
Size: 10,000 acres **Closest Town:** Mountain View, WY   P ⬛ ▲ ⵜ

45 Ptarmigan Loop

Description: Long trail loop into High Uintas Wilderness in the Wasatch-Cache National Forest. Coniferous forest, riparian meadow and shrub, and high alpine vegetation. This is the only white-tailed ptarmigan habitat in Utah. Moose, elk, mule deer, and various small mammals also occupy alpine basins during the summer.

Viewing Information: Moderate to limited probability of viewing white-tailed ptarmigan from July through September. The species is brown with white belly, wings, and tail in summer; it is pure white except for black eyes and bill during the winter. Closely aligned with alpine willow habitats. VIEWING SITE IS A MINIMUM NINE- TO TWELVE-MILE HIKE (ONE-WAY). SUMMER STORMS ARE FREQUENT AND TRAIL IS ARDUOUS.

Directions: Take Interstate 80 east into southwest Wyoming and exit at Fort Bridger. Turn south on Wyoming 414 to Mountain View. Then take Wyoming 410 west then south for approximately seven miles. Continue south on gravel surface Forest Road 072, then turn east about four miles past the Forest Boundary onto Forest Road 017. Continue for 5.5 miles to Forest Road 077, then turn south and travel another 3.5 miles to Henry's Fork Trailhead. Hike from here on Forest Trail 117 nine miles to ptarmigan habitat. For more viewing, continue another three miles through Gunsight Pass to Painter Basin (Forest Trail 068) on the Ashley National Forest.

Ownership: USFS (307 782-6555)
Size: Eleven-mile hike **Closest Town:** Mountain View, WY   P ⬛ ▲ ⵜ 🚶

46 | Whitney Basin

Description: Scenic national forest area with alpine fir, aspen, sagebrush, grass-forb meadows, and willow and sedge riparian habitats. Classic moose/beaver/northern harrier habitat in Mill City Creek, Meadow Creek, and West Fork of Bear Creek. Considerable raptor viewing in the Moffet Basin area during fall migration. Horned larks, white-crowned sparrows, and pine siskins are among the wide array of bird species present from spring through late summer.

Viewing Information: Moderate to limited probability of viewing moose from mid-June through October. Beaver viewing is limited, but best toward evening near active beaver dams. Songbirds and other species are best seen during June and early July. Raptor viewing probability is high during September from mid-day until evening along the west side of Whitney Basin; use binoculars here.

Directions: *From Utah 150, turn west at the Whitney Reservoir Road (Forest Road 032). Travel seven miles to Forest Road 071 and EITHER turn south to the Whitney Reservoir area (Forest Roads 032 and 069) OR continue west on Forest Road 071 to Moffet Basin area. An alternate viewing site is Gold Hill Road (Forest Road 109) just 2.6 miles west of Utah 150. This road (Forest Roads 081 and 160) may be taken south about three miles for moose/beaver/harrier viewing. THIS A DIRT ROAD AND SHOULD BE AVOIDED WHEN WET.*

Ownership: USFS (307 789-3194)
Size: 22,000 acres **Closest Town:** Evanston, WY   **P** ⚏ 🚶

47 | Bald Mountain

Description: Short, high-elevation trail with scenic vista in the Wasatch-Cache National Forest that leads to mountain goat habitat. Trail also passes through pika habitat. Other small mammals and bird species may be seen here.

Viewing Information: Moderate to limited probability of viewing goats; moderate for pika. Trail traverses the southwest side and top of the Bald Mountain, nearly 12,000 feet in elevation. EXERCISE CAUTION. HIKE SHOULD BE AVOIDED DURING LIGHTNING STORMS.

Directions: *From Kamas, drive east on Utah 150 approximately 29 miles. Turn north at Bald Mountain Trailhead sign and drive 1/4 mile to parking area.*

Ownership: USFS (783-4338)
Size: 1,200 acres **Closest Town:** Kamas   **P** 🪧 🚶

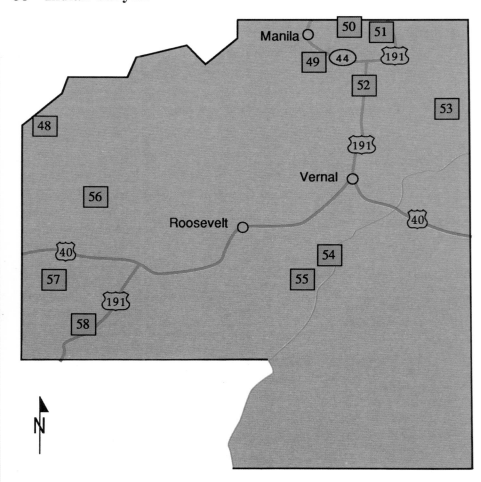

48 | Mirror Lake Nature Trail

Description: Interpretive nature trail around high mountain lake (10,200-foot elevation) in extremely scenic setting. Alpine forest habitats support red crossbill, pine grosbeak, pine marten, and Canada lynx as well as the more common Clark's nutcracker, Canada jay, chipmunk, dark-eyed junco, white-crowned sparrow, and swallows.

Viewing Information: Moderate probability of viewing common species, limited for crossbills and grosbeaks, rare for marten and lynx. Viewing season from mid-June through September. Interpretive center with viewing telescope planned for 1990.

Directions: From Kamas, travel east on Utah 150 for 31.3 miles. Turn east at the Mirror Lake Campground sign and drive .25 mile.

Ownership: USFS (783-4338)
Size: 50 acres **Closest Town:** Kamas   P𝕒𝕟⛺▲&🚶$

49 | Sheep Creek

Description: A loop road along rugged and scenic Sheep Creek Canyon (known as the Sheep Creek Geological Loop). In the fall, view spawning kokanee salmon. View elk, mule deer, moose, and bighorn sheep during the spring and early summer. A variety of small mammals including the yellow-bellied marmot, chipmunk, and cottontail rabbit may be seen in the summer. Raptors may be observed in the spring through fall.

Viewing Information: High probability of viewing spawning kokanee during September. This unique, disease-free, early-run kokanee strain is the source of eggs for hatchery systems in Utah and Wyoming. MOVE QUIETLY ALONG THE STREAM BANK. TO PROTECT SPAWNING SALMON AND THEIR EGGS, DO NOT ENTER THE STREAM OR THROW ANYTHING INTO IT . Probability of viewing elk and deer is high in the spring and early summer, while probability of viewing moose and bighorn is limited to moderate in the spring through fall. Day use by the public is permitted. HOWEVER, OVERNIGHT CAMPING IS PROHIBITED IN THE CANYON FROM MAY 15 TO SEPTEMBER 30 DUE TO THE HIGH RISK OF FLASH FLOODS.

Directions: From Manila, travel south on Utah 44 for seven miles to Sheep Creek. Turn west onto Sheep Creek Geological Loop (Forest Road 218) and continue to the intersection with Utah 44. The road is paved.

Ownership: USFS (784-3445)
Size: 14-mile drive **Closest Town:** Manila   P𝕒𝕟⛺&🚶

Description: A boat trip on the reservoir within Flaming Gorge National Recreation Area that offers an excellent opportunity to view osprey, peregrine falcon, and bighorn sheep in scenic, cliff habitat. Swallows, swifts, gulls, grebes, and other bird species may also be viewed.

Viewing Information: High probability of viewing osprey at the numerous nest sites identified on the reservoir. These birds nest atop rocky pinnacles adjacent to the water and may be viewed in Horseshoe Canyon. Peregrine falcon (a federally listed endangered species) nest on cliffs above the lake; probability of viewing is limited due to their relative scarcity. Bighorn sheep may sometimes be viewed by boat on Kingfisher Island and between Skull Creek and Hideout Canyon on the north side of the lake during spring and early summer. Reservoir has visitor center and visitor contact stations.

Directions: Boats may be launched on Flaming Gorge Reservoir from several developed ramps: Lucerne, Cedar Springs, Mustang Ridge, Antelope Flats, and Sheep Creek Bay.

Ownership: USFS (784-3445)
Size: 375 miles of shoreline
Closest Town: Dutch John or Manila

The majestic beauty of Flaming Gorge is an inviting setting to view the variety of wildlife species which live in the surrounding habitats. JACK A. RENSEL

51	Lucerne Peninsula

Description: Sagebrush peninsula on northwest side of Flaming Gorge Reservoir that features pronghorn antelope, prairie dogs, and other small mammals. Wetland areas adjacent to the reservoir support ducks, geese, herons, egrets, grebes, and other species.

Viewing Information: High probability of viewing antelope year-round along the entrance road and in the Lucerne Valley Campground. Unique opportunity to view and photograph antelope at very close range in campground. High probability of viewing waterfowl and wading birds in Linwood Bay, south of the road. Visitor contact station at campground.

Directions: From Manila, travel east on Utah 43 for about four miles to the Lucerne Valley turnoff. Turn southeast onto Forest Road 146 and drive about four miles to the campground. Roads are paved.

Ownership: USFS (784-3445)
Size: 5,000 acres **Closest Town:** Manila P 🏕🛶 ⛩ ▲ ♿ $ 🚤

The pronghorn antelope is unique to North America. It has true horns, but like the antlers of the deer family, the horn sheath is shed each year. Lucerne Peninsula is one of the best places in Utah to view antelope at close range. CHRISTOPHER CAUBLE

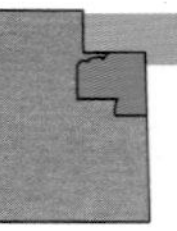

52 East Uinta Mountains Drive

Description: Beautiful drive through conifer, aspen, juniper, mountain brush, sagebrush, and small meadow habitats in the Ashley National Forest. View moose, elk, mule deer, bighorn sheep, marmots, and birds of prey at various times throughout the year. Cottontail and white-tailed jackrabbits, golden and bald eagles, chipmunks, and squirrels may be viewed while driving or by stopping at turnouts.

Viewing Information: Limited probability of viewing moose, which may be seen in small groups or individually at the higher elevations in the coniferous forest. Elk and mule deer are visible in relatively large numbers from December through March on low-elevation winter ranges north of Vernal and near Manila; also view from April through August on higher elevation spring and summer ranges. Viewing probability for these species decreases from high to limited as the summer lengthens. Probability of viewing bighorn sheep is moderate to limited during the spring and summer months along the switchbacks south of Sheep Creek. Moderate probability of viewing raptors overhead in open areas along the drive route and from Red Canyon Visitor Center. The route has been designated a National Scenic Byway; a portion is within the Flaming Gorge National Recreation Area. Key viewing features include the Red Canyon Visitor Center, Moose Pond, Sheep Creek Bay, McKee Draw, and the Chevron Resources overlook. The route also has geologic interpretive signs.

***Directions:** From Vernal, drive north on U.S. 191 to the junction with Utah 44. Route continues west and north on Utah 44 to the town of Manila.*

Ownership: USFS (789-1181)
Size: 64-mile drive **Closest Town:** Vernal or Manila P $

The yellowbelly marmot is common to rocky habitats throughout Utah. Its characteristic high-pitched chirp gives a warning of danger to other marmots. JAN L. WASSINK

53 Diamond Mountain

Description: Sage grouse strutting grounds on typical sage brush/meadow habitat. Spectacular breeding display by male grouse. Mule deer and elk may also be observed in the vicinity.

Viewing Information: High probability of viewing sage grouse on their strutting grounds from late March though mid-April. View birds on private lands both north and south of roadway. Binoculars are essential here. VIEW BIRDS ONLY FROM ROADSIDE TURNOUTS—LEAVING THE COUNTY ROAD WOULD DISRUPT BREEDING ACTIVITY. LANDS ARE PRIVATE; HONOR THE RIGHTS OF PRIVATE LANDOWNERS AND DO NOT TRESPASS.

Directions: In Vernal, turn east at U.S. 191 and 500 North. Follow paved county road and signs toward Jones Hole Fish Hatchery about 26 miles (between Jackson Draw and Crouse Reservoir signs) to the viewing site. Sage grouse may be observed strutting on the north side of the road between Diamond Gulch and the Pot Creek turnoff.

Ownership: PVT (UDWR 789-3103)
Size: 80 acres **Closest Town:** Vernal

54 Ouray National Wildlife Refuge

Description: Refuge along 12 miles of the Green River featuring cottonwood and willow habitats, marshes, and desert shrub and grassland types. View waterfowl and shorebird migrations in the spring and fall, waterfowl and shorebird broods in the summer, bald and golden eagle concentrations in late winter, and mule deer populations year-round. Over 2,000 nesting pairs of 14 duck species occur here, and some 209 bird species use the refuge.

Viewing Information: Tremendous viewing area. High probability of viewing all species in appropriate seasons. Spring waterfowl and shorebird migration peaks in April; the fall migration peaks in October. Facilities include visitor contact station, observation tower, interpretive brochures, and a ten-mile auto tour on gravel roads.

Directions: From Vernal, travel about 14 miles west on U.S. 40 and turn south onto Utah 88. Travel south another 16 miles to refuge headquarters.

Ownership: USFWS (789-0351)
Size: 11,483 acres **Closest Town:** Vernal **P**

 The scientific name of the yellowbelly marmot — *Marmota flaviventris*—**translates as "the yellowbelly mountain mouse."**

55 | Pariette Wetlands

Description: A unique marsh complex surrounded by many miles of arid desert, featuring freshwater ponds, alkali bulrush, diverse emergent vegetation, and wet meadow types. Mallard, gadwall, cinnamon teal, pintail, and Canada geese are the most common waterfowl species. Herons, egrets, white-faced ibis, and American bittern are common wading bird species. The site was developed in 1972 to improve waterfowl production and provide seasonal habitat for other species including ring-necked pheasant, mourning dove, sandhill and whooping cranes, and peregrine falcon. A wide variety of raptors including the bald eagle, harrier, and prairie falcon also use the area.

Viewing Information: High probability of viewing wetland species in the spring through fall. Interpretive signs and brochures are available.

Directions: *Take U.S. 40 to Fort Duchesne, turn south, and drive about five miles (just past the Duchesne River). At the Myton "Y", turn south off the road to Myton onto the dirt road and travel another 16 miles across Leland Bench to Pariette Wash. Follow signs to overlook. THESE ARE GRAVEL AND DIRT ROADS AND SHOULD BE AVOIDED WHEN WET. CONTACT THE BLM FOR MAPS AND ROAD CONDITIONS.*

Ownership: BLM (789-1362)
Size: 9,033 acres **Closest Town:** Roosevelt **P**

56 | Yellowpine Trail

Description: Two trails from Yellowpine Campground that follow Rock Creek downstream from Upper Stillwater Reservoir. One trail lies between the reservoir and the campground. Following construction in 1990, a second trail for the physically challenged will be at the campground. Featured species include birds of forest habitats, plus mule deer, elk, moose, marmots, and other small mammals.

Viewing Information: Moderate probability of viewing all species during May through October. Hiking trail may be accessed by vehicle from either the reservoir or campground end. Trail for the physically challenged will feature cassette tape tours and interpretive pamphlet.

Directions: *Take U.S. 40 to Duchesne and turn north onto Utah 87. Drive 16.5 miles to the Mountain Home turnoff, then drive 2.9 miles to Mountain Home. There, turn west on Forest Road 134 (Rock Creek Road) and drive 21.2 miles to Yellowpine Campground. All roads are paved.*

Ownership: USFS (738-2482)
Size: Five-mile hike, 1/4-mile physically challenged trail
Closest Town: Altamont   P⛺🚻▲♿🚶$

57 Strawberry River Wildlife Management Area

Description: Beautiful riparian corridor and stream flanked by high cliffs. Cottonwood, willow, plus sedge and grass meadows are the primary habitats. Wide variety of warblers, vireos, swallows, jays, woodpeckers, thrushes, waterfowl, and raptors. Yellowbelly marmots, beaver, squirrels, chipmunks, and mule deer may also be observed.

Viewing Information: High probability of viewing featured species during May through October. Use turnouts and binoculars. View waterfowl, raptors, and deer in the winter, although conditions may preclude easy access. Brown trout may be viewed spawning during October in the riffles if they are approached quietly.

Directions: Take U.S. 40 to approximately three miles east of Fruitland and turn south onto the Red Creek Road. Continue 6.3 miles to the Camelot Resort sign, then turn west. Go past the Resort and private campground to begin route. Wildlife Management Area is signed. Travel up-canyon about eight miles to private property gates. ROAD FROM U.S. 40 IS A DIRT ROAD AND SHOULD BE AVOIDED WHEN WET. THIS ROAD MAY ALSO BECOME IMPASSABLE DURING WINTER SNOW CONDITIONS.

Ownership: UDWR (789-3103)
Size: Eight-mile drive **Closest Town:** Duchesne P ☂ 🚶

58 Indian Canyon

Description: Elk and mule deer winter range along highway. Sagebrush, grassland, and juniper woodland habitats.

Viewing Information: High to moderate viewing probability from January through March. Site offers opportunity to view these species at close distances (50 to 200 yards). PLEASE DO NOT LEAVE PARKING AREAS TO MINIMIZE STRESS PLACED ON WINTERING ANIMALS.

Directions: From Duchesne, drive 16 miles south on U.S. 191. Viewing site is 2.5 miles south of national forest boundary on east side of highway. HIGHWAY MAY BE ICY AND SNOWPACKED DURING VIEWING PERIOD.

Ownership: USFS (738-2482)
Size: 100 acres **Closest Town:** Duchesne P

 Utah has 10 resident species of owls, including the burrowing owl—the only owl that uses abandoned rodent burrows for nesting and cover.

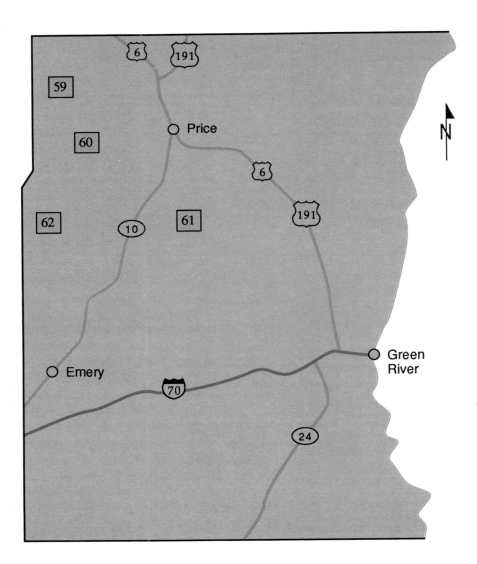

59 | **North Skyline Drive**

Description: Spectacular 30-mile drive that follows the divide between the Great Basin and the Colorado River drainage. Scenic view from 9,000-foot elevation. North aspect conifer, aspen, sagebrush, and grassland vegetative communities. Wide variety of wildlife including mule deer and elk in spring and early summer, and migrating raptors in early fall. Raptor species include Cooper's and sharp-shinned hawks, goshawk, red-tailed hawk, golden eagle, and others.

Viewing Information: High probability of viewing elk and mule deer in the spring, especially during morning and evening. Fall raptor migration begins in late August and runs through November; viewing probability is high mid-day to evening.

Directions: Take U.S. 6 to about eight miles west of Soldier Summit and turn south at Tucker Rest Area. Travel along ridgetop to Utah 31, the end of the viewing route. Skyline Drive intersects Utah 31 about eight miles east of Fairview. Gravel surface road, except the FIRST THREE MILES SOUTH OF TUCKER REST AREA IS A DIRT ROAD WHICH SHOULD BE AVOIDED WHEN WET.

Ownership: USFS (637-2817)
Size: 30-mile drive **Closest Town:** Fairview P 🏕 🚶 ⛐ ▲

Although native to Utah, elk were eliminated from most of their natural habitat by the turn of the century. Today, under careful management, the species flourishes in much of the state. This elk calf will retain its spots until fall. MICHAEL S. SAMPLE

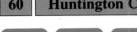

60 | **Huntington Canyon**

Description: Highly scenic drive along riparian corridor that meanders through mixed aspen-conifer habitats. Viewing species include mule deer, songbirds, beaver and its habitat, and a variety of other mammals.

Viewing Information: Viewing probability is moderate. Drive-by site for spring, summer, and fall viewing, with numerous turnouts and overlooks. Wide variety of songbirds use the cottonwoods, aspen, and willow types in the riparian corridor. Small groups of mule deer best seen in mornings and evenings. Tie Fork Canyon (approximately four miles up-canyon from the Manti-LaSal National Forest boundary on the east side) is a botanical peculiarity with 22 tree species.

Directions: Begin on Utah 31 (a National Scenic Byway) 12 miles west of Huntington or 18 miles east of Fairview on Utah 31. Drive route runs between the National Forest boundary on the east to Electric Lake on the west. PAVED STATE HIGHWAY IS CLOSED OCCASIONALLY BY WINTER STORMS.

Ownership: USFS (637-2817)
Size: 16-mile drive **Closest Town:** Huntington   P 🅿 🚶 🛆 ⛺

Due to its inherent high-level productivity and diversity, riparian habitat is the most important habitat type for a majority of Utah wildlife. Huntington Canyon, shown here, demonstrates these qualities. JIM COLE

61 | Desert Lake Waterfowl Management Area

Description: Marsh habitat in unique desert setting—the largest managed wetland complex in southeastern Utah. The marsh is a typical bulrush, cattail, saltgrass complex, while the upland desert is commonly Indian ricegrass, shadscale, and snakeweed. Area features waterfowl, shorebirds, wading birds, upland species, and a variety of songbirds and raptors. Mule deer, beaver, and prairie dogs also use the site.

Viewing Information: High probability of viewing wetland species during the spring and summer. Public is confined to county road between April 1 and August 15 to protect breeding waterfowl and shorebirds. Dogs are prohibited during the spring-summer period. Group tours are available and may be coordinated in advance. Bird list is available.

Directions: From Price, travel south on Utah 10, turn southeast onto Utah 155, and drive about two miles to Elmo turnoff. Drive east to Elmo, then travel one mile east and turn south at the "Dinosaur Quarry" sign. Travel 0.7 miles to the entrance of the Waterfowl Management Area. Drive south, then east an additional 2.3 miles through the marsh to an overlook. ROADS ARE DIRT AND GRAVEL SURFACE, AND SHOULD BE AVOIDED WHEN WET.

Ownership: UDWR (637-3310)
Size: 3,000 acres **Closest Town:** Huntington **P**

62 | Joes Valley Ski Trail

Description: Two two-mile cross-country ski trail loops adjacent to Joes Valley Reservoir in mule deer winter range. Ski track is set as needed. Mule deer, bald eagle, snowshoe hare, and occasionally weasel can be seen.

Viewing Information: Moderate probability of viewing deer; small groups can be viewed in pinyon-juniper and ponderosa pine communities from December to mid-March. Hundreds of mule deer can be seen along highway en route to site.

Directions: From Orangeville, travel west on Utah 29 for approximately 15 miles to the west side of Joes Valley Reservoir. Ski trailhead sign is .5 mile north of Joes Valley Campground turnoff at Littles Creek. Blue diamond markers designate trail location and direction of travel. Snow conditions can vary; call ahead for current conditions.

Ownership: USFS (384-2372)
Size: Four-mile ski trail **Closest Town:** Orangeville **P**

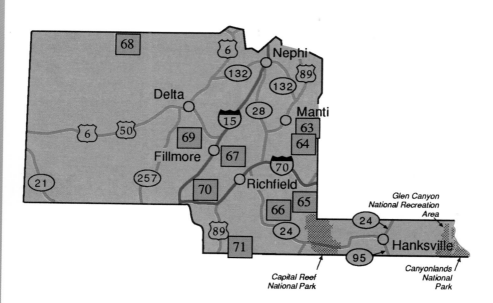

63 Elks Knoll

Description: A unique natural area little-used or disturbed by man located in central Utah featuring grassland, mountain brush, and coniferous and aspen forest habitats. Typical Wasatch Plateau species are found here including coyote, badger, snowshoe hare, blue grouse, elk, mule deer, meadowlark, and red-tailed hawk.

Viewing Information: Viewing probability is moderate to limited during the June to September viewing period. No motor vehicle access to the site; IT IS A STEEP HIKE AND THERE ARE NO TRAILS. The site is managed by the Manti-LaSal National Forest.

Directions: From Ephraim drive southeast up Ephraim Canyon about 3.5 miles to Manti-LaSal National Forest boundary. Continue east on Forest Road 115 (gravel surface) to the Skyline Drive (Forest Road 139). Turn south and continue approximately seven miles, then turn north on primitive road and continue about one mile. Hike west from this point (the east side of Elks Knoll). FOREST ROAD 139 IS A DIRT ROAD AND IS EXTREMELY SLIPPERY WHEN WET—AVOID DURING WET PERIODS.

Ownership: USFS (283-4151)
Size: 40 acres **Closest Town:** Ephraim

64 Ferron Reservoir Interpretive Trail

Description: Nature trail through scenic montane basin with coniferous forest, riparian meadow and shrub, and aspen vegetative communities. Observe a variety of high elevation forest species including beaver, marmot, ground squirrel, northern harrier, American kestrel, red-tailed hawk, snowshoe hare, many songbirds, and occasionally mule deer and elk. Trail to be constructed and signed by late 1990 viewing season.

Viewing Information: Moderate probability of viewing all featured species during July through September. Interpretation of relationships between resident wildlife species and their respective habitats is provided along the trail.

Directions: From Ferron, turn west off Utah 10 at U.S. Forest Service office onto Canyon Road. Continue about six miles to the National Forest boundary and beginning of Forest Road 022. Drive on gravel and dirt roads another 22 miles to Ferron Reservoir. Trail is located near National Forest Campground at reservoir inlet. DRIVING ON FOREST ROAD 22 MAY BE DIFFICULT WHEN WET OR SNOWPACKED. CONTACT FOREST SERVICE OFFICE IN FERRON FOR CURRENT CONDITIONS.

Ownership: USFS (384-2372)
Size: One-half mile hike **Closest Town:** Ferron

65 **Hogan Pass**

Description: Classic Great Basin mule deer and elk habitat with expansive sagebrush, aspen, and grassland types. View up to 400 elk and a few bald eagles during the winter. See elk, mule deer, American kestrel, northern harrier, red-tailed hawk, whitetail jackrabbit, sage grouse, and mountain bluebird in the spring and early summer. A Utah prairie dog colony is located about 1/4 mile southwest of the Pass.

Viewing Information: Moderate probability of viewing all featured species during the appropriate season. Use binoculars here, especially during the winter.

Directions: Just north of Loa, turn onto Utah 72 and drive 9.4 miles northeast through Fremont to the national forest boundary. Viewing drive begins here and continues about nine miles to Hogan Pass.

Ownership: USFS (836-2800)
Size: Nine-mile drive **Closest Town:** Loa

Shiras moose are surprisingly common in Utah, and their range is expanding. Fish Lake Basin is one of several viewing areas where moose may be observed.
GLENN VAN NIMWEGEN

66 | Fish Lake Basin

Description: Extremely scenic National Forest site with diverse habitat including riparian meadow and shrub, aspen, conifer, and grassland types. Site chosen for Utah prairie dog re-establishment (two colonies). View waterfowl, shorebirds, and wading birds in the marsh habitats. Raptors here include northern harrier, American kestrel, red-tailed hawk, osprey, and golden eagle. View mule deer, elk, and moose in the spring through fall. Small mammals seen occasionally include badger, ground squirrel, red squirrel, marmot, and weasel. Spawning trout may be observed near the Twin Creek parking lot from April to June.

Viewing Information: A May through October viewing site. High probability of viewing Utah prairie dog. Moderate to limited probability for most other species in the appropriate season. Best viewing from turnouts. Self-guided auto tour and evening nature programs presented on weekend evenings from June through August.

Directions: Turn off Utah 24 onto Utah 25 at Fish Lake turnoff. Continue to Fish Lake Recreation Area and begin viewing. Continue to the northern end of the Basin, about ten miles.

Ownership: USFS (836-2811)
Size: 8,000 acres **Closest Town:** Loa

The threatened Utah prairie dog may be viewed at Fish Lake Basin. Long term recovery programs for this species will take place in the West Desert, Paunsaugunt Plateau, and the Owapa Plateau. FRED HIRSCHMANN

67 **Chalk Creek**

Description: Fishlake National Forest drive into narrow, scenic, riparian canyon. Drive passes through oakbrush and juniper woodland and into cottonwood, willow, and rocky cliff riparian zone. View a variety of arid land and riparian-associated bird species including scrub and Steller's jay, green-tailed towhee, violet-green swallow, yellow warbler, and others. Mule deer are common during morning and evenings in oakbrush zone. Various species of raptors, snakes, and lizards are also common.

Viewing Information: Moderate probability of viewing featured species from May through October. Use binoculars to scan cliff faces for nesting swallows and other species. The bobcat may occasionally be viewed here. Its scientific name, *Lynx rufus*, literally translates as "the reddish wildcat with shining eyes."

Directions: In Fillmore turn east off Main Street at 200 South (Canyon Road). Continue about 2.5 miles to the national forest boundary and begin viewing from Forest road 100. Travel another 6.7 miles to the end of the tour at the Pistol Rock Picnic area. THE ROAD IS GRAVEL SURFACE, SOMEWHAT STEEP AND NARROW, AND IS COMMONLY WASHBOARD. RECREATION VEHICLES ARE NOT RECOMMENDED.

Ownership: USFS (743-5721)
Size: Seven-mile drive **Closest Town:** Fillmore   P 🏠 ⛽ ⛺ ♿ 🚶

A yellow warbler feeds insects to its young. Like the many species of warblers which breed in North America, this species selects shrub or woodland habitat commonly associated with riparian vegetation. JAN L. WASSINK

| 68 | **Fish Springs National Wildlife Refuge** |

Description: A wetland oasis on the south edge of the Great Salt Lake Desert. About 10,000 acres of saline spring-fed marsh forms the habitat base. Swans, geese, many species of ducks, herons, egrets, avocets, stilts, and grebes may be observed here. A variety of mammals including kit fox, coyote, badger, long-tailed weasel, and blacktail jackrabbit use the desert habitat that surrounds the marsh. Eleven species of reptiles may also be found here.

Viewing Information: High probability of viewing wetland bird species from spring through fall; winter viewing is also possible. Total numbers are lower during the spring, but species diversity is greater. Thousands of waterfowl use the site in the fall. A loop auto tour with brochures is available. Check with refuge headquarters for any seasonal restrictions.

Directions: From Delta, travel northwest on U.S. 6 about 10 miles, then turn onto Utah 174 and travel 42 miles to the end of the improved road. Continue north to northwest another 14.6 miles to the Pony Express Road, then turn west and travel 6.7 miles to the refuge headquarters. Or, exit Utah 199 at Dugway and travel Old Pony Express Route for about 61 miles THROUGH UNINHABITED DESERT ON DIRT ROADS. AVOID THESE ROADS WHEN WET. Or, from Wendover, Nevada, travel south on U.S. 93 to Ibapah turnoff. Travel on both gravel and dirt roads southeast to Callao, then to Fish Springs. THE VIEWING SITE IS REMOTE! THE NEAREST GAS STATION IS 40 MILES FROM REFUGE.

Ownership: USFWS (831-5353)
Size: 17,992 acres **Closest Town:** Delta   P 🏕🏕

The kit fox inhabits sandy desert and juniper woodland areas in Utah. It is the smallest dog-like mammal in the state and can be readily identified by its size and extremely large ears. JACK A. RENSEL

69 | Clear Lake Waterfowl Management Area

Description: Extensive marsh area in west desert. Open water, cattail, bulrush, and wet meadow types provide habitat for numerous wetland species. Observe many duck species, as well as avocets, stilts, egrets, herons, coots, northern harriers, curlews, and a variety of songbirds. Coyotes and foxes may also be viewed occasionally.

Viewing Information: High probability of viewing wetland bird species in the spring through fall; migrational peaks in April and November are best. Access by vehicle on county road through the area; otherwise access is limited to hiking approximately ten miles of dikes. ACCESS PROHIBITED IN WATER OR VEGETATED AREAS. Site has viewing platform .25 mile off road. Self-contained camping is permitted adjacent to parking areas.

Directions: *From Delta, follow U.S. 6 west for about five miles, then turn south onto Utah 257. Continue south for 15.7 miles and turn east at Clear Lake Waterfowl Management Area sign. Drive about six miles on gravel surface road to the Clear Lake Area and begin viewing. Continue east two miles to end of viewing.*

Ownership: UDWR (864-2924)
Size: 6,150 acres **Closest Town:** Delta or Fillmore **P**

70 | Fremont Indian State Park

Description: State park in scenic canyon adjacent to Interstate 70. Mule deer winter range in sagebrush and juniper woodland habitat. Rock outcrops form canyon walls; look for various snakes, lizards, and small mammals including marmots and cottontail rabbits in the vicinity of the outcrops. Cottonwood and willow riparian zone along Clear Creek. Songbirds and beaver may be found in riparian zones. Bald eagles also may be viewed during the winter.

Viewing Information: High probability of viewing mule deer from December through April. Moderate probability of observing other species during spring and early summer. Visitor center interprets Fremont Indian lifestyle including key role of four bird species (bald eagle, red-tailed hawk, northern shrike, cliff swallow) involved in Hopi legend.

Directions: *Turn off Interstate 70 at exit 15 south of Richfield and follow the signs.*

Ownership: UDPR (527-4631)
Size: 1,108 acres **Closest Town:** Richfield P⌂♨▲♿⚡$

71 | Otter Creek Reservoir

Description: High desert reservoir that is habitat for wetland bird species including common loons, pelicans, cormorants, grebes, gulls, terns, coots, ducks, and geese. Ospreys are common summer residents; golden and bald eagles, and rough-legged hawks may be viewed during the winter.

Viewing Information: High probability of viewing featured species in relative abundance during appropriate periods. Best wetland bird viewing during spring and fall migrations. Binoculars are essential since the vantage may be long in some cases. State Park facilities at south end of reservoir. The Park is a fee area.

Directions: From U.S. 89, take Utah 62 east to the reservoir, then proceed north on Utah 62 to mile marker 17, and then take dirt road to reservoir overlook. DIRT ROAD IS SLIPPERY WHEN WET. AVOID USE DURING THIS PERIOD. Small boats may be launched from Otter Creek State Park ramp (fee area).

Ownership: UDPR/BLM (UDPR 624-3268)
Size: 3,200 acres **Closest Town:** Antimony

The magnificent great blue heron can be found in all water habitats in Utah. In flight, it can be distinguished from cranes by its habit of folding its neck back on its shoulders. RALEIGH MEADE

72 **Big Flat**
73 **Parowan Front**
74 **Pine Valley**
75 **Snow Canyon State Park**
76 **Lytle Ranch Preserve**
77 **Joshua Tree Natural Area**
78 **Zion National Park**
79 **Tom Best Loop**
80 **Escalante State Park**
81 **Aquarius Plateau**
82 **Boulder Mountain**
83 **Henry Mountains**

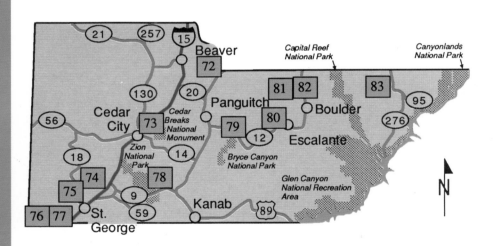

72 | **Big Flat**

Description: Extensive grassland meadow in coniferous forest on ridgetop in the Tushar Mountains. Perhaps the state's best summer mule deer viewing opportunity, with 50 to 200 deer visible in the meadows. Raptors are prominent here, particularly during fall migration. Coyotes are common, their howls often heard at night.

Viewing Information: High probability of viewing mule deer from June through August. A very large complement of mature bucks may be observed. Although a wild population, deer are relatively tame making this an excellent photography opportunity. Elk, hawks, American kestrels, golden eagles, mountain bluebirds, western meadowlarks, horned larks, and numerous grassland sparrow species offer moderate viewing probability.

Directions: From Beaver, take Utah 153 east for about 23 miles and begin viewing. Continue south about four miles to Gunsight Flat and end of tour. Continue east about 13 miles to U.S. 89 at Junction. Highway is paved from Beaver east about 18 miles. The remainder of the route is gravel or dirt roads. AVOID DIRT ROADS WHEN WET. ACCESS FROM THE JUNCTION SIDE IS STEEP, NARROW, AND WINDING—NOT RECOMMENDED FOR LONG TRAILERS.

Ownership: USFS (438-2436)
Size: 800 acres   **Closest Town:** Beaver or Junction   **P**

Mule deer may be viewed throughout the summer in many areas of Utah, but the Big Flat site offers easy viewing of large numbers, including many mature bucks.
MICHAEL S. SAMPLE

73 | Parowan Front

Description: Winter mule deer range providing roadside viewing of hundreds of deer. Typical deer winter habitat consisting of juniper woodland and sagebrush types.

Viewing Information: High probability of viewing mule deer from December through early April. View from vehicle. MOTORIZED VEHICLES ARE LIMITED TO EXISTING ROADS AND TRAILS FROM JANUARY 1 THROUGH APRIL 30 TO PROTECT THIS IMPORTANT WINTER HABITAT.

Directions: Drive the frontage road adjacent to Interstate 15 (east side of I-15) between Cedar City and Summit. Road is gravel surface.

Ownership: BLM/UDWR (BLM 586-2458)
Size: Eight-mile drive **Closest Town:** Cedar City

74 | Pine Valley

Description: Small, picturesque valley in the Pine Valley Mountains, featuring meadowlands surrounded by ponderosa pine forest. Mountain brush and juniper woodland on west side of valley. Observe a large concentration of mule deer in the spring through fall. Jays, woodpeckers, songbirds, cottontail rabbits, red squirrels, and chipmunks are common forest residents. American kestrels, northern harriers, and other hawks are also common.

Viewing Information: High probability of viewing mule deer from April through October. Moderate probability of viewing all other species.

Directions: From St. George, drive north on Utah 18 for 25 miles to Central and turn east onto Forest Road 035. Begin viewing at the Dixie National Forest boundary. Drive 6.5 miles into Pine Valley and continue east another 2.3 miles to the east end of Ponderosa Campground and end of tour.

Ownership: USFS (574-2949)
Size: Nine-mile drive **Closest Town:** Pine Valley   P 🏕 ⛽ 🅾 ⛰ ♿ $ 🚶

The ringtail is a rarely seen nocturnal species which may be found in dry, rocky, cliff habitats as far north as the Ogden River.

75 | Snow Canyon State Park

Description: Scenic canyon in desert setting, which includes some geologically significant sites. Featured reptiles include desert tortoise, gila monster, and other lizards and snakes. Songbirds, including a variety of hummingbird species, are abundant during the breeding season.

Viewing Information: THE DESERT TORTOISE IS A FEDERALLY LISTED THREATENED SPECIES. DO NOT COLLECT, HANDLE, OR MOLEST IN ANY WAY. Limited to rare probability of viewing tortoise and gila monster; high to moderate for other species. Viewing season is spring through fall. Park features a self-guided nature trail and an interpretive display at the campground where native flora and fauna are displayed. RATTLESNAKES ARE COMMON HERE—USE CAUTION WHEN WALKING THROUGH AREA.

Directions: From St. George, drive north about 11 miles on Utah 18 and follow the signs.

Ownership: UDPR (628-2255)
Size: 6,500 acres **Closest Town:** St. George   P🏠🪟⛺🅿️▲♿💲🚶

76 | Lytle Ranch Preserve

Description: Remote extension of Mojave desert in extreme southwestern Utah. Riparian intrusion in Joshua tree, cholla, cactus, and creosote bush vegetation. Bird species unique to Utah include vermilion flycatcher, phainopepla, white-winged dove, Gambel's quail, crissal thrasher, verdin, cactus wren, Costa's hummingbird, and summer tanager. A variety of rattlesnake and lizard species, as well as the desert tortoise, reside here. Look for beaver in riparian habitat.

Viewing Information: THE DESERT TORTOISE IS A FEDERALLY LISTED THREATENED SPECIES. DO NOT COLLECT, HANDLE, OR MOLEST IN ANY WAY. Probability of viewing most species is high in spring (March through May), decreases in summer, and increases in fall (October through December), although site may be visited year-round. Tours for larger groups (25 or more) may be arranged through Brigham Young University who operates the preserve. RATTLESNAKES ARE COMMON HERE—USE CAUTION WHEN WALKING THROUGH AREA.

Directions: From St. George, travel north on Utah 18. Turn west toward Santa Clara and continue to Shivwits. Follow old highway southwest about 11.8 miles to signed turnoff (west) to Lytle Ranch. Travel 10.8 miles on gravel surface road to Lytle Ranch headquarters.

Ownership: NP (378-5052)
Size: 462 acres **Closest Town:** Santa Clara   P🏠⛺▲

77 | **Joshua Tree Natural Area**

Description: A transition zone between Great Basin and Mojave Desert vegetation. Large Joshua trees, barrel cactus, cholla, and creosote bush dominate the landscape. Desert tortoise is the featured viewing species, although a number of desert-dwelling birds, snakes, lizards, kit foxes, and small mammals may also be viewed here.

Viewing Information: DESERT TORTOISE IS A FEDERALLY LISTED THREATENED SPECIES. DO NOT COLLECT, HANDLE, OR MOLEST IN ANY WAY. Limited to rare probability of viewing tortoise, but April and May are best. Interpretive signs and a brochure for the area are planned for 1990. RATTLESNAKES ARE COMMON HERE—USE CAUTION WHEN WALKING THROUGH AREA.

Directions: *Follow Interstate 15 into Arizona and take the Littlefield exit. Drive north on paved road 10 miles, turn east at "Woodbury-Hardy Desert Study Area" sign and travel two miles on a gravel road to the Joshua Tree Natural Area.*

Ownership: BLM (673-4654)
Size: 3,040 acres **Closest Town:** Littlefield, AZ

In Utah the threatened desert tortoise occurs only in the extreme southwestern corner of the state. It constructs underground burrows for shelter and lays as many as fifteen eggs annually. J. KIRK GARDNER

78 | Zion National Park

Description: Colorful canyon and mesa scenery which includes erosion and rock-fault patterns. Zion Canyon of the North Fork of the Virgin River is a lush oasis in the harsh, desert environment which provides habitat for a variety of species. The diversity of habitat is enhanced by the 5000-foot elevational change within the park. Mule deer are common, and bird species include golden eagle, dipper, pygmy owl, turkey, three species of nuthatch, and many more. The rare peregrine falcon and Mexican spotted owl also inhabit the park. Beaver, antelope ground squirrel, ringtail, and porcupine also may be observed here. Other species include the canyon tree frog and the king snake.

Viewing Information: Viewing probability for mule deer, beaver, porcupine, and a wide variety of songbirds is moderate to high. Species such as the peregrine falcon, Mexican spotted owl, and ringtail cat are rare. The park has 30 miles of scenic drives and over 100 miles of trails, which have varying degrees of difficulty. Many of the trails are self-guided tours. There are two visitor centers where wildlife and other natural history publications are available.

Directions: Park headquarters and the main visitor center are located on Utah 9 about one mile east of Springdale. The Kolob Canyons entrance is east off Interstate 15 at Exit 40. The park may be reached from the east by turning west off U.S. 89 at Mt. Carmel and driving about 13 miles to the east park entrance. DUE TO LARGE VISITOR NUMBERS AND THE LIMITED ROAD SYSTEM, LARGE VEHICLES ENTERING THROUGH THE PARK'S EAST OR SOUTH ENTRANCES MAY BE SUBJECT TO SPECIAL RESTRICTIONS AND CHARGES. CONTACT ENTRANCE STATIONS FOR SPECIFIC INFORMATION.

Ownership: NPS (772-3256)
Size: 147,000 acres **Closest Town:** Springdale

Zion National Park contains spectacular and abundant wildlife. The park's 30 miles of scenic drives and more than 100 miles of trails provide excellent wildlife-viewing opportunities.
MICHAEL S. SAMPLE

79 | Tom Best Loop

Description: A loop drive that extends through a high desert shrub community and into a ponderosa pine forest. Area is elk winter range and year-round pronghorn antelope habitat. Various species of raptors can be seen throughout the year.

Viewing Information: Moderate probability of viewing all species in appropriate seasons. Elk congregate here from several summer ranges (including Mt. Dutton, Paunsaugunt, Aquarius Plateau, and Boulder Mountain) and may be observed from late fall through early spring. Elk viewing may be limited to the Utah 12 portion of the loop when winter storms close other portions. View pronghorn and raptors from anywhere along the drive.

Directions: Begin at junction of Utah 12 and Utah 22, travel north 11.6 miles, and turn west onto Forest Road 117. Continue west and then south to Utah 12 (Tom Best Spring sign). Drive east to the point of origin at the Utah 22 intersection. All segments of the loop are gravel or dirt roads except Utah 12, which is paved. AVOID DIRT ROADS WHEN WET.

Ownership: USFS/BLM (USFS 676-8815)
Size: 34-mile drive **Closest Town:** Panguitch

80 | Escalante State Park

Description: Reservoir area in desert canyon that is one of the few wetland bird viewing sites in southern Utah. Sagebrush and juniper woodland habitats, with willow and cottonwood riparian vegetation at the reservoir margin. Wide variety of ducks, coots, grebes, herons, and swallows. Eagles, osprey, American kestrel, and other hawks and falcons may be observed here. Cottontail and blacktail jackrabbits, Uinta ground squirrels, and beaver are also common.

Viewing Information: High probability of viewing all waterfowl, wading birds, and shorebird species from spring though fall. Moderate to limited probability of viewing small and hoofed mammals; high to moderate for raptors. Binoculars essential here. Park features 1.5-mile trail through petrified forest and cultural prehistoric sites.

Directions: From Escalante, drive two miles west on Utah 12. Turn north to State Park.

Ownership: UDPR (826-4466)
Size: 1,351 acres **Closest Town:** Escalante   P ⌂ 𝔸 ☂ ⛺ $ 𝆏

81 Aquarius Plateau

Description: Unique, high elevation pronghorn antelope viewing. Extensive stands of sagebrush openings in coniferous and aspen forests. Observe elk, mule deer, and occasionally black bear. American kestrels, red-tailed hawks, and golden eagles are common, as are mountain bluebirds and other songbirds. Prairie dogs may be observed at northern end of drive.

Viewing Information: High probability of viewing antelope and moderate probability of viewing all other species from mid-May to November. Binoculars essential for best view.

Directions: *Turn off Utah 12 about 4.2 miles west of Escalante onto Main Canyon Road. Continue northwesterly for 5.3 miles to the intersection of Forest Road 140 and turn north here. Continue 31.5 miles to the intersection with Forest Road 154 and turn north. Follow Forest Road 154 for 20 miles north to the intersection with Pine Creek Road and either continue 11.2 miles north to Loa or turn northeasterly onto Pine Creek Road and travel about 6 miles to Utah 24, some 2.5 miles south of Bicknell. Entire route is gravel or dirt roads. DIRT ROADS, SUCH AT THE MAIN CANYON ROAD, SHOULD BE AVOIDED WHEN WET.*

Ownership: USFS (826-4221 or 425-3702)
Size: 70-mile drive **Closest Town:** Escalante or Bicknell   P 禾 ▲ 𝑘

82 Boulder Mountain

Description: Highly scenic Dixie National Forest drive through conifer, aspen, mountain brush, and grassland vegetative types. Turkey, mule deer, a wide variety of songbirds, squirrels, chipmunks, snowshoe hares, and cottontail rabbits may be seen here.

Viewing Information: Rare probability of viewing turkey; moderate to high for the other species. Best turkey viewing is between Garkane Power Plant road and Pleasant Creek Campground. Viewing season is spring through fall.

Directions: *Paved drive route on Utah 12 between Boulder and Torrey. USE CAUTION WHEN PARKING AND/OR PULLING OUT ALONG HIGHWAY.*

Ownership: USFS (425-3702 or 826-4221)
Size: 32-mile drive **Closest Town:** Boulder   P ▲ 𝑘

 The literal translation of the buffalo's scientific name *(Bison bison)* **is "the humpbacked ox that stinks during the rutting season."**

83 | Henry Mountains

Description: Large desert mountain range with peaks more than 11,000 feet high. Desert shrub, juniper woodland, grassland, conifer, and aspen communities dominate the landscape. Primary viewing attraction is the bison. This is one of the few free-roaming bison herds in the country, with approximately 200 adult animals. Mule deer, a variety of raptors, small mammals, and songbirds also may be viewed here.

Viewing Information: Moderate probability of viewing bison year-round. View at higher elevations in spring and summer, and at lower elevations on the west side of the Henry's in fall and winter.

Directions: MAPS AND INFORMATION AVAILABLE AT BLM OFFICE IN HANKSVILLE. In Hanksville, turn south off Utah 24 at 100 East and follow signs toward the Lonesome Beaver Recreation Area. Drive about 13 miles and continue south on the right fork of a "Y" junction. Drive another 11.5 miles and turn west to Bull Creek Pass (begin bison tour on spring/summer habitat). Continue over the Pass another five miles and turn west toward McMillan Spring Campground. Travel west six miles past the campground and continue west at the intersection (continue tour in fall/winter habitat). Follow signs to Utah 24. Continue another 7.5 miles, then take the northerly road just past Sweetwater Wash (end bison tour). Leave area via Notum Road and travel an additional 23.5 miles north to Utah 24 approximately 27 miles west of Hanksville. ALL ROADS ARE DIRT AND CROSS NUMEROUS WASHES. SOME ROAD SEGMENTS ARE QUITE STEEP. AVOID AREA DURING OR FOLLOWING HIGH-INTENSITY RAINSTORMS. FOUR WHEEL-DRIVE VEHICLES STRONGLY RECOMMENDED.

Ownership: BLM (542-3461)
Size: 230,000 acres; 94-mile drive **Closest Town:** Hanksville P 🏠 ⊼ ▲ 🚶

The Henry Mountains are home to some 200 buffalo. Bison were reestablished in Utah in 1941 and have occupied the Henry's since 1963. JIM COLE

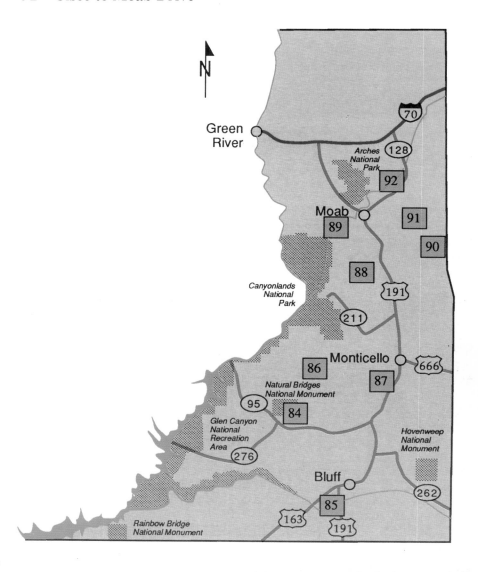

| 84 | **Natural Bridges National Monument** |

Description: Eight-mile loop road through scenic, natural bridge country. Vegetation is largely pinyon-juniper woodland. Featured species include midget-faded rattlesnake and a variety of other reptiles, mammals, and songbirds.

Viewing Information: Limited to rare probability of viewing rattlesnake, but interpretive information in visitor center will familiarize visitors with the species. High probability of viewing swifts and swallows during spring and summer.

Directions: From Blanding, travel south on U.S. 191 for four miles, then travel west on Utah 95 for about 34 miles, then turn off onto Utah 275 and travel four miles to visitor center.

Ownership: NPS (259-5174)
Size: 7,400 acres **Closest Town:** Blanding P🚻🏕️⛰️$🚶

| 85 | **San Juan River** |

Description: Two-day river float trip on San Juan River. Songbirds and waterfowl are best viewed during the spring and fall migrations; peregrine falcon and other raptors may be observed in spring and summer. Desert bighorn can generally be viewed along the river's south side.

Viewing Information: High probability of viewing most species in spring and early summer. Limited to rare probability of viewing desert bighorn and peregrine falcon due to their relatively low population numbers and expansive habitat. Binoculars are essential on this trip. Float the river at any time, but winter is generally too cold. Float permits required—obtain through BLM.

Directions: Put-in at Sand Wash approximately three miles west of Bluff; take-out at Mexican Hat. For guided tours, a river guide list is available through the BLM's Moab District.

Ownership: BLM (587-2141)
Size: 27-mile float **Closest Town:** Bluff P🚻🏕️⛰️🛶$

The riparian zones in Utah's deserts serve as migration corridors for a number of bird species that spend the winter in South America, such as the yellow warbler and western tanager.

Utah is home to both the mountain cottontail and the desert cottontail. Both have home ranges of less than fifteen acres and usually favor habitats of brushy thickets. This young cottontail rests beneath lupine flowers. MICHAEL S. SAMPLE

86 | Elk Ridge

Description: Open ponderosa pine stands with excellent diversity of wildlife species. Aspen and mountain brush habitats are also well distributed here. Mule deer, elk, cottontail rabbits, and a variety of songbirds are prominent. Turkey and black bear may occasionally be viewed. Waterfowl and shorebirds nest at Duck Lake.

Viewing Information: High probability of viewing mule deer, Abert squirrels, cottontail rabbits, and songbirds during the late spring to early fall. Limited probability of viewing American kestrel, goshawk, and red-tailed hawk; turkey and black bear are rare.

Directions: From Blanding, take U.S. 191 south to Utah 95. Turn west and travel about 5.5 miles to South Cottonwood Creek road (Forest Road 106). Drive about 25 miles to the intersection with Forest Road 095, then travel west 2.5 miles to Forest Road 088. Travel north or south along Elk Ridge for additional viewing opportunities. FOREST ROAD 106 HAS SEVERAL STREAM FORDS WHICH MAY BE IMPASSABLE AT HIGH WATER AND AFTER RAINSTORMS. AVOID THESE DIRT ROADS WHEN WET.

Ownership: USFS (587-2041)
Size: 220,000 acres **Closest Town:** Blanding   P ⚊

87 | Devil's Canyon Campground

Description: Scenic ponderosa pine and mountain brush habitat for Abert squirrel. Look for its bulky nests high in ponderosa pines. Also view a variety of other small mammals and birds here.

Viewing Information: Moderate probability of viewing Abert Squirrel; best viewing is in the spring through fall. View at Manti-LaSal National Forest campground or from a hike up Bulldog Canyon.

Directions: From Blanding, travel eight miles north on Utah 191. Turn west at Devil's Canyon Campground sign and continue approximately .25 miles to the campground.

Ownership: USFS (587-2041)
Size: 56 acres **Closest Town:** Blanding   P ⚊ ⚊ ⚊ ⚊ $

 The Abert squirrel builds a large, rather bulky nest high in ponderosa pines. In Utah the species is found only east of the Colorado River.

| 88 | **Canyon Rims Recreation Area** |

Description: Recreation area with sagebrush plateau surrounded by red-rock desert canyons. Pronghorn antelope is the most prominent area resident; jackrabbits, other small mammals, and raptors are also common.

Viewing Information: Swallows, swifts, and raptors are commonly observed at scenic canyon overlooks located at Needles (end of pavement), Anticline (end of gravel road), and Canyonlands (reached only by 4-WD) overlooks. Moderate probability of viewing antelope.

Directions: From Monticello, travel approximately 17 miles north on U.S. 191. Turn west into Canyon Rims Recreation Area.

Ownership: BLM (259-6111)
Size: 85,000 acres **Closest Town:** Monticello P🏠 ⪥ ▲ 🚶

| 89 | **Dead Horse Point State Park** |

Description: Spectacular vista overlooking deep red-rock desert canyons and Colorado River. Typical desert bighorn sheep habitat. Also view a variety of other species including raptors, ground squirrels, cottontail rabbits, songbirds, lizards, and horned toads.

Viewing Information: Rare probability of viewing desert bighorn sheep. Visitor center at Park headquarters plans desert bighorn interpretive display. Binoculars required to see most species.

Directions: From Moab, follow U.S. 191 north for nine miles, then turn west onto Utah 313. Travel 23 miles to visitor center.

Ownership: UDPR (259-6511)
Size: 5,082 acres **Closest Town:** Moab P🏠 ⪥ ▲ 🚶 $

The yellow-headed collared lizard is one of the unique species of reptiles which may be viewed in the southeastern Utah desert types. See this species at the LaSal Loop viewing site. TIM CLARK

85

| **90** | **Old LaSal** |

Description: Grassland/farmland site with opportunity to view 30 to 50 elk during early spring. Animals are visible in the morning and evening.

Viewing Information: High probability of viewing elk in season; best during the early vegetation green-up period (March through early April). Binoculars are useful here. Site is on private land; HONOR THE RIGHTS OF PRIVATE LANDOWNERS AND VIEW WILDLIFE FROM PUBLIC ROADWAYS ONLY. Please use caution when parking along roadways.

Directions: From U.S. 191 (between Moab and Monticello), turn east onto Utah 46. Travel about 16 miles to the old town site of Old LaSal. Viewing is north of highway. Also turn north on Forest Road 072, marked by sign to Burro Pass Trail. Viewing opportunity for about .5 mile on this gravel road.

Ownership: PVT (UDWR 637-3310)
Size: 500 acres **Closest Town:** LaSal

| **91** | **LaSal Loop** |

Description: Highly scenic national forest loop road that passes through desert shrub, mountain brush, juniper woodland, riparian, aspen, and conifer habitats. Mule deer are abundant and elk may occasionally be viewed along route. Side trips from loop offer opportunities to view pika and, in another case, varied species of reptiles.

Viewing Information: High probability of viewing mule deer along loop road and pikas along trails leading from Warner Campground. Also high probability of viewing collared, mountain short-horned, and/or northern side-blotched lizards along red rock face of western national forest boundary on Brumley Ridge. EXERCISE CAUTION AT BRUMLEY CREEK—RATTLESNAKES ALSO DWELL HERE.

Directions: From Moab, follow U.S. 191 south for about eight miles, then turn east and begin loop drive. Continue east, then north for about 37 miles to intersection with Utah 128, where loop ends. Reach Brumley Creek lizard viewing area by turning east two miles past the start of loop route onto the Ken's Lake road (gravel surface). Continue east about 2.5 miles to the viewing site at the national forest boundary. Reach LaSal Mountain pika viewing site by traveling east on the loop road for approximately 15 miles and turning east at the Warner Lake road (Forest Road 067). Travel 5.4 miles to Warner Campground (THIS SECTION IS DIRT ROAD—AVOID WHEN WET), then take either Burro Pass Trail or Miners Basin Trail until you reach talus slopes (rock slide areas). Begin viewing and hearing pikas.

Ownership: USFS (259-7155)
Size: 37-mile drive **Closest Town:** Moab

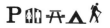

92 | Cisco to Moab Drive

Description: Spectacularly scenic drive along the Colorado River featuring riparian corridor in red rock desert. Year-round viewing of waterfowl, herons, and egrets. View bald eagles in the winter, roosting in cottonwoods. Observe prairie dogs at the Cisco end of the drive during spring and summer. Mule deer are common and desert bighorn sheep may occasionally be observed along the river's north side.

Viewing Information: High probability of viewing waterfowl, wading birds, and raptors in appropriate seasons. Desert bighorn viewing is rare. HIGHWAY MAY BE ICY IN WINTER. ALSO USE CAUTION WHEN USING HIGHWAY TURNOUTS. Colorado River can be floated from Fisher Towers to Moab.

Directions: Drive route begins on Utah 128 two miles north of Moab, and continues on Utah 128 to two miles west of Cisco.

Ownership: BLM (259-6111)
Size: 46-mile drive **Closest Town:** Moab

Desert bighorn sheep inhabit dry, cliff areas in parts of southeastern Utah. Patience, binoculars, and luck are needed to view these rare animals. RON SANFORD

PLEASE HELP IMPROVE THE NEXT GUIDE

Since this guide is the first of its kind in Utah, the sponsors welcome comments concerning your experiences. Observations about the selection of sites, management of the areas, adequacy of the facilities, need for interpretive information, usefulness of the format, and site directions will all be considered when this guide is revised in the future. Site evaluation forms are available upon request should you wish to recommend additional areas. Send letters to Defenders of Wildlife, 333 South State Street, Suite 173, Lake Oswego, OR 97034. Your comments will be shared with the sponsors.

MORE BOOKS FROM FALCON PRESS

Falcon Press publishes a wide variety of outdoor books and calendars, including the state-by-state series of wildlife viewing guides called the Watchable Wildlife Series. If you liked this book, please look for the companion books on other states.

If you want to know more about outdoor recreation in Utah, then look for *The Hiker's Guide to Utah*. Falcon Press also publishes guidebooks on hiking, fishing, scenic driving, river floating, and rockhounding for most other western states.

To purchase any of these books, please check with your local bookstore or call toll-free 1-800-582-BOOK. When you call, please ask for a free catalog listing all the fine books and calendars from Falcon Press.

Falcon Press Publishing Co., Inc., P.O. Box 1718, Helena, MT 59624